Dimensions Math®
Textbook 3A

Authors and Reviewers

Bill Jackson

Jenny Kempe

Cassandra Turner

Allison Coates

Tricia Salerno

Consultant

Dr. Richard Askey

Singapore Math Inc.

Published by Singapore Math Inc.

19535 SW 129th Avenue
Tualatin, OR 97062
www.singaporemath.com

Dimensions Math® Textbook 3A
ISBN 978-1-947226-08-1

First published 2018
Reprinted 2019, 2020

Printed in China

Acknowledgments

Editing by the Singapore Math Inc. team.
Design and illustration by Cameron Wray with Carli Fronius.

Preface

The Dimensions Math® Pre-Kindergarten to Grade 5 series is based on the pedagogy and methodology of math education in Singapore. The curriculum develops concepts in increasing levels of abstraction, emphasizing the three pedagogical stages: Concrete, Pictorial, and Abstract. Each topic is introduced, then thoughtfully developed through the use of problem solving, student discourse, and opportunities for mastery of skills.

Features and Lesson Components

Students work through the lessons with the help of five friends: Emma, Alex, Sofia, Dion, and Mei. The characters appear throughout the series and help students develop metacognitive reasoning through questions, hints, and ideas.

The colored boxes   and blank lines in the textbook lessons are used to facilitate student discussion. Rather than writing in the textbooks, students can use whiteboards or notebooks to record their ideas, methods, and solutions.

Chapter Opener

Each chapter begins with an engaging scenario that stimulates student curiosity in new concepts. This scenario also provides teachers an opportunity to review skills.

Think

Students, with guidance from teachers, solve a problem using a variety of methods.

Learn

One or more solutions to the problem in **Think** are presented, along with definitions and other information to consolidate the concepts introduced in **Think**.

Do

A variety of practice problems allow teachers to lead discussion or encourage independent mastery. These activities solidify and deepen student understanding of the concepts.

Exercise

A pencil icon ━━━━━━━━▶ at the end of the lesson links to additional practice problems in the workbook.

Practice

Periodic practice provides teachers with opportunities for consolidation, remediation, and assessment.

Review

Cumulative reviews provide ongoing practice of concepts and skills.

Emma   Alex   Sofia   Dion   Mei

Contents

Chapter	Lesson	Page

Chapter		Lesson	Page

Chapter		Lesson	Page

Chapter 1

Numbers to 10,000

How can we count all of these straws easily without losing track of how many we have counted?

Think

How many straws are there?

Learn

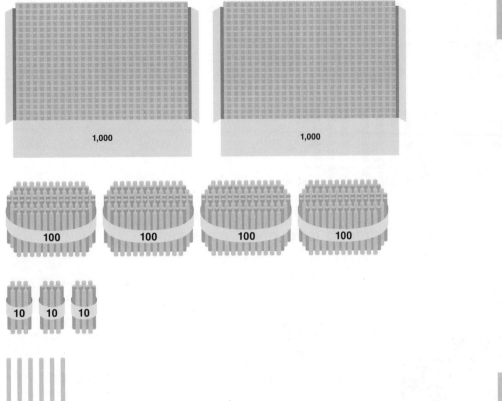

two thousand, four hundred thirty-six

We usually put a comma after the digit in the thousands place.

2 thousands, 4 hundreds, 3 tens, and 6 ones make 2,436.

2,000 + 400 + 30 + 6 = ▭

Do

1 (a)

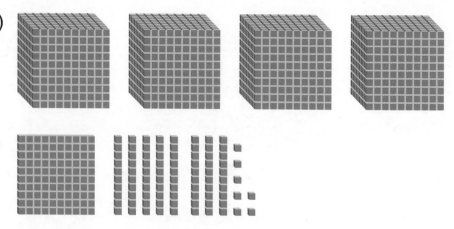

4 thousands, 1 hundred, 8 tens, and 7 ones make [   ].

4,000 + 100 + 80 + 7 = [   ]

4,0 0 0
1 0 0
8 0
7
$\rightarrow$ 4,1 8 7

four thousand, one hundred eighty-seven

(b)

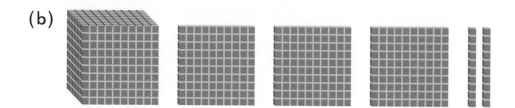

1 thousand, 3 hundreds, and 2 tens make [   ].

1,000 + 300 + 20 = [   ]

1,0 0 0
3 0 0
2 0
$\rightarrow$ 1,3 2 0

one thousand, three hundred twenty

(c)

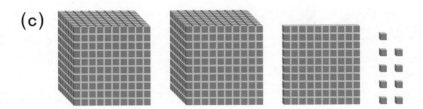

2 thousands, 1 hundred, and 9 ones make ▢ .

2,000 + 100 + 9 = ▢

2,0 0 0 → 2,1 0 9
  1 0 0
      9

two thousand, one hundred nine

(d)

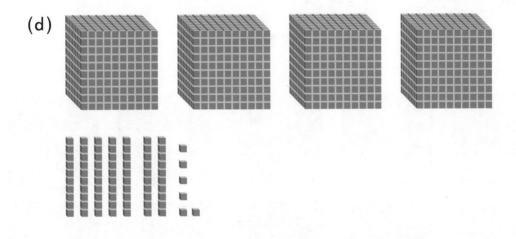

4 thousands, 7 tens, and 6 ones make ▢ .

4,000 + 70 + 6 = ▢

4,0 0 0 → 4,0 7 6
    7 0
        6

four thousand, seventy-six

(e)

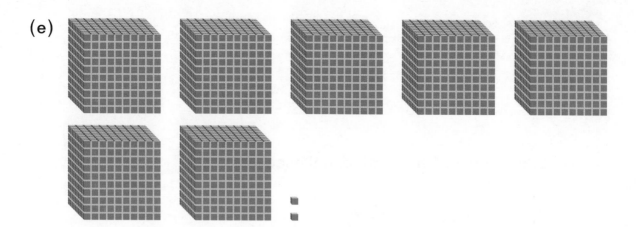

7 thousands and 2 ones make [ ].

7,000 + 2 = [ ]

7, 0 0 0 → 7, 0 0 2
       2

seven thousand, two

❷

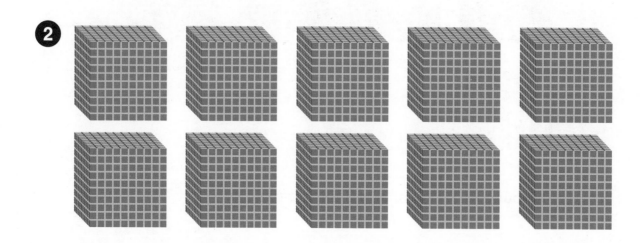

10 thousands is [ ].

1 0, 0 0 0

ten thousand

3 Read each number.
Show each number with place-value cards.
Write each number in words.

(a) 1,299   (b) 9,494   (c) 7,250

(d) 2,306   (e) 5,027   (f) 6,005

4 (a) $3,000 + 300 + 40 + 9 =$ ▢

(b) $7,000 + 80 + 1 =$ ▢

(c) $9,000 + 600 + 4 =$ ▢

(d) $5,000 + 3 =$ ▢

(e) $5 + 90 + 400 + 6,000 =$ ▢

5 Count on by ones from 4,987 to 5,011.

6 Count on by ones and write the missing numbers.

(a) 7,007   7,008   7,009   ▢ ▢ ▢ ▢

(b) 6,996   6,997   6,998   ▢ ▢ ▢ ▢

Think

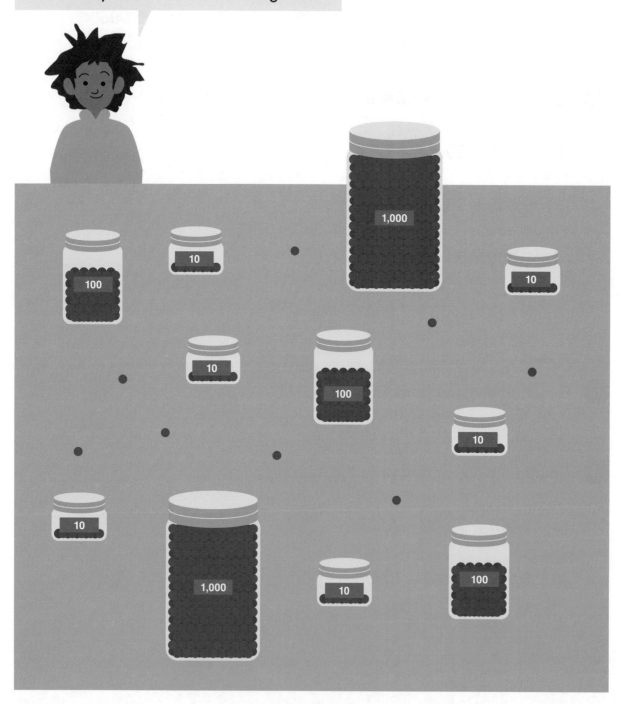

How many beads do I have altogether?

Learn

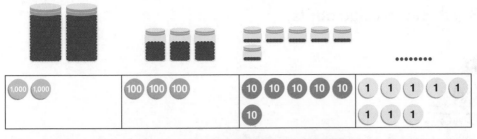

Thousands	Hundreds	Tens	Ones
2	3	6	8

2,000 + 300 + 60 + 8 = ⬚

Dion has ⬚ beads.

2,0 0 0

The digit 2 in 2,368 is in the thousands place.
It stands for 2 thousands.
Its value is 2,000.

3 0 0

The digit 3 in 2,368 is in the hundreds place.
It stands for 3 hundreds.
Its value is ⬚.

6 0

The digit 6 in 2,368 is in the tens place.
It stands for ⬚ tens.
Its value is 60.

8

The digit 8 in 2,368 is in the ones place.
It stands for 8 _____.
Its value is ⬚.

2,3 6 8

Do

1 Show 4,573 with place-value cards.

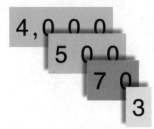

(a) The digit 4 in 4,573 is in the _____ place.

(b) The digit 5 in 4,573 stands for 5 _____.

(c) The digit 7 in 4,573 stands for 7 _____.

(d) The digit 3 in 4,573 is in the _____ place.

 2

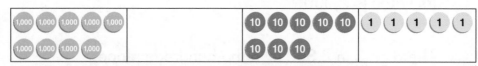

Thousands	Hundreds	Tens	Ones
9	0	8	5

(a) The digit ⬜ in 9,085 is in the ones place.

(b) The digit 0 in 9,085 is in the _____ place. Its value is 0.

(c) The value of the digit 9 in 9,085 is ⬜.

(d) The digit ⬜ in 9,085 stands for ⬜ tens.

3 Write the number.

(a)
| 1,000 1,000 1,000 1,000 1,000 | 100 100 100 100 | 10 10 10 | 1 1 1 1 1 |
| 1,000 1,000 | | | 1 1 1 1 |

(b)
| 1,000 1,000 1,000 1,000 1,000 | 100 100 100 100 100 | | 1 1 1 1 |
| 1,000 1,000 1,000 | 100 | | |

(c)
| 1,000 1,000 1,000 1,000 1,000 | 100 100 100 100 100 10 | | |
| | 100 100 100 100 | | |

4 (a) Write the number in words.

| 4,982 | 2,308 | 9,250 | 5,029 |

(b) In what place is the digit 2 in each number, and what is its value?

5 (a) $6,069 = 6,000 + \boxed{} + 9$

(b) $7,402 = 7,000 + 400 + \boxed{}$

(c) $5,300 = \boxed{} + 300$

(d) $5,008 = 5,000 + \boxed{}$

(e) $1,953 = \boxed{} + 1,000 + 3 + 50$

(f) $8,808 = 8 + \boxed{} + 8,000$

Exercise 2 • page 4

Think

(a) Show the number 2,300 with place-value discs.

(b) How can Dion show this number using only 100s?

I only have 100s.

(c) How can Emma show this number using only 10s?

I only have 10s.

Learn

(a)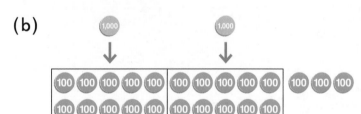

(b)

1,000 = 10 hundreds

2,000 = ⬚ hundreds

300 = ⬚ hundreds

2,300 = ⬚ hundreds

23 hundreds = 2,300

(c)

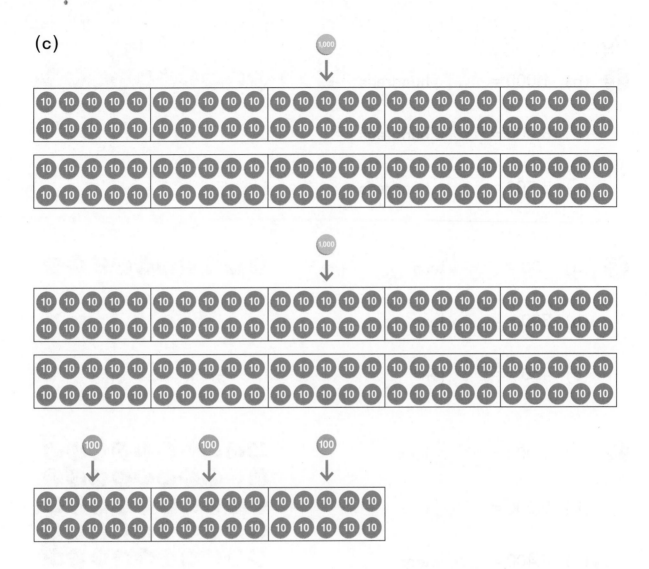

1,000 = 100 tens

2,000 = ▭ tens

300 = ▭ tens

2,300 = ▭ tens

200 tens + 30 tens = 230 tens
230 tens = 2,30**0**

Do

1 (a) 1,000 = [____] hundreds

(b) 6,000 = [____] hundreds

(c) 6,700 = [____] hundreds

2 (a) 100 = [____] tens

(b) 300 = [____] tens

(c) 450 = [____] tens

3 (a) 1,000 = [____] tens

(b) 5,000 = [____] tens

(c) 5,400 = [____] tens

(d) 5,420 = [____] tens

4 (a) How many hundreds make 4,900?

(b) How many hundreds make 7,200?

(c) How many tens make 5,600?

(d) How many tens make 9,210?

5 What number is made up of...

(a) 80 hundreds?

(b) 52 hundreds?

(c) 700 tens?

(d) 380 tens?

6 (a) 9,560 is ⬜ hundreds and ⬜ tens.

(b) 4,320 is ⬜ thousands and ⬜ tens.

(c) 2,047 is ⬜ tens and ⬜ ones.

7 (a) 42 hundreds and 4 ones make ⬜ .

(b) 8 thousands, 20 tens, and 3 ones make ⬜ .

Exercise 3 • page 7

Lesson 4
Comparing Numbers

(4)

Think

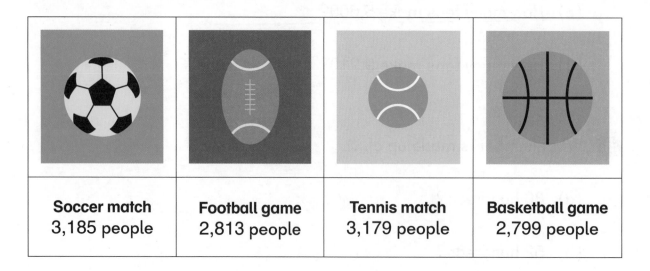

| Soccer match 3,185 people | Football game 2,813 people | Tennis match 3,179 people | Basketball game 2,799 people |

Which game at the sports center had the greatest attendance?
Which had the least?

Learn

Football game
2,813 people

Soccer match
3,185 people

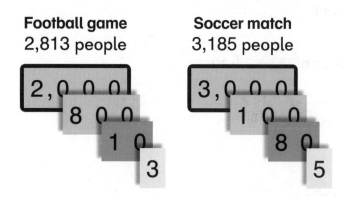

Compare the digits in the greatest place first.

2 thousands is less that 3 thousands.
2,813 < 3,185

There were fewer people at the _____ than the _____.

Basketball game
2,799 people

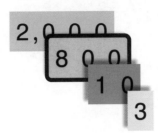

Football game
2,813 people

The digits in the thousands place are the same. Compare the digits in the hundreds place.

7 hundreds is less than 8 hundreds.

2,799 < 2,813

There were fewer people at the _____ than the _____.

Soccer match
3,185 people

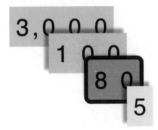

Tennis match
3,179 people

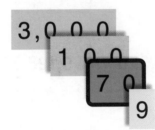

The digits in the thousands place and the hundreds place are the same. Compare the digits in the tens place.

8 tens is more than 7 tens.

3,185 > 3,179

There were more people at the _____ than the _____.

The _____ had the most people.

The _____ had the least people.

Do

1 What sign, > or <, goes in the ◯?

(a) 6,999 ◯ 7,010

(b) 3,567 ◯ 3,559

(c) 7,085 ◯ 7,084

(d) 2,001 ◯ 898

2 Arrange the numbers in order from least to greatest.

(a)

| 2,387 | 1,960 | 3,002 | 998 |

(b)

| 5,258 | 5,099 | 5,320 | 5,198 |

(c)

| 7,569 | 6,567 | 6,559 | 6,570 |

3 Arrange the numbers in order from greatest to least.

(a)

| 8,067 | 8,068 | 8,065 | 8,066 |

(b)

| 4,132 | 4,321 | 4,123 | 4,312 |

4 What sign, >, <, or =, goes in the ◯?

(a) 7,000 + 800 + 50 + 9 ◯ 7,959

(b) 8,205 ◯ 8,000 + 100 + 90 + 6

(c) 5,000 + 200 + 50 ◯ 5,250

(d) 2,000 + 60 + 9 ◯ 2,000 + 100

5 What are the greatest and least 4-digit numbers that you can make using all the digits for each set of numbers...

(a) 1, 9, 5, and 4

(b) 7, 0, 3, and 8   0,837 is not a 4-digit number.

(c) 2, 4, 7, and 4

(d) 9, 0, 4, and 0

(e) 7, 6, 7, and 7

6 (a) What is the greatest 4-digit number?

(b) What is the least 4-digit number?

Exercise 4 • page 10

Think

What is the distance between each tick mark on each **number line**?
What number is indicated by each letter?

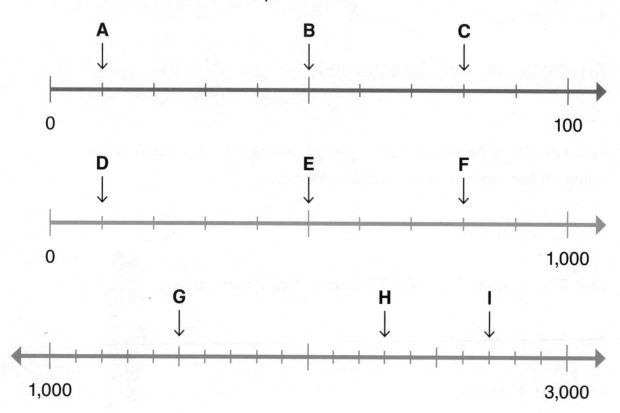

The numbers on the number lines increase from left to right.

Learn

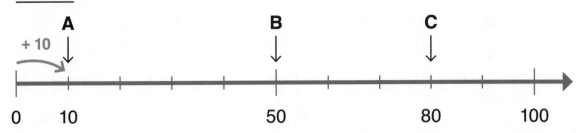

The distance between each tick mark on the blue number line is 10.

50 is halfway between 0 and 100.

How many increments of 10 are there between 0 and 100?

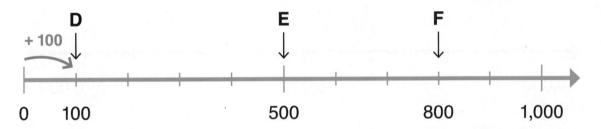

The distance between each tick mark on the green number line is ⬜.

⬜ is halfway between 0 and 1,000.

How many increments of 100 are there between 0 and 1,000?

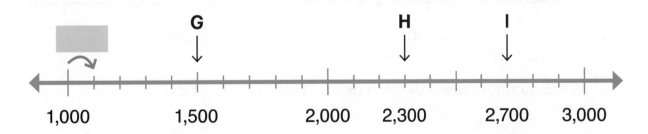

The distance between each tick mark on the red number line is ⬜.

⬜ is halfway between 1,000 and 2,000.

1

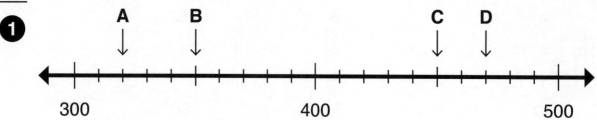

(a) What numbers are indicated by each letter, A, B, C, and D?
(b) Which number is the greatest?
(c) Which number is the least?

2

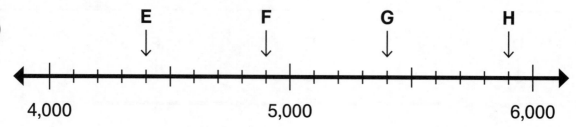

(a) What numbers are indicated by each letter, E, F, G, and H?
(b) Which number is the greatest?
(c) Which number is the least?

3

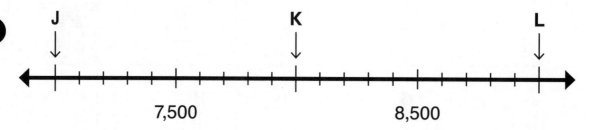

What numbers are indicated by each letter, J, K, and L?

4

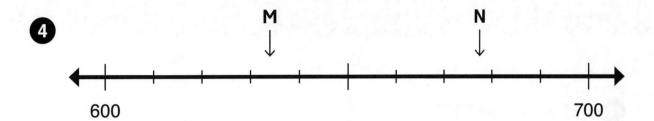

600   700

(a) M is between 630 and 640.
It is nearer to the tick mark for ▮▮▮.

(b) N is nearer to the tick mark for ▮▮▮.

I think M is 632.

5

P   Q

2,000   3,000

(a) P is halfway between the tick marks for ▮▮▮ and ▮▮▮.
P indicates the number ▮▮▮.

(b) Q is nearest to the tick mark for ▮▮▮.

6

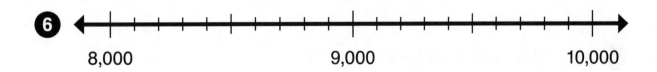

8,000   9,000   10,000

Where are these numbers on the number line above?

(a) 8,200   (b) 8,500
(c) 8,550   (d) 9,510
(e) 9,700   (f) 9,790

Exercise 5 • page 13

1 In the number 5,273...

(a) The digit [ ] is in the hundreds place.

(b) The digit [ ] is in the ones place.

(c) The digit 5 is in the _____ place.

(d) The digit 7 stands for 7 _____.

2 (a) Read each number.

| 3,820 | 8,207 | 2,008 | 5,982 |

(b) What is the value of digit 8 in each number?

(c) In what place is the digit 2 in each number?

3 (a) 2,000 + 500 + 20 + 8 = [ ]

(b) 3,000 + 9 = [ ]

(c) 7,000 + [ ] + 50 = 7,250

(d) 40 + 2 + 6,000 = [ ]

4 The height of the Burj Khalifa skyscraper
in Dubai is about 2,722 ft.
Write this number in words.

5 (a) 5,000 =  hundreds

(b) 8,600 = [ ] hundreds

(c) 2,000 = [ ] tens

(d) 1,340 = [ ] tens

6 Write the number.

(a) 4 thousands, 4 hundreds, 3 tens, and 7 ones

(b) 7 hundreds and 5 thousands

(c) 65 hundreds

(d) 806 tens

7 What sign, > or <, goes in the ◯?

(a) 4,987 ◯ 5,002

(b) 5,438 ◯ 5,356

(c) 3,125 ◯ 3,139

(d) 804 ◯ 8,004

(e) 8,000 + 50 + 2 ◯ 8,060

(f) 2,808 ◯ 8 + 80 + 2,000

8 Arrange the numbers in order from least to greatest.

| 6,560 | 6,559 | 6,752 | 5,759 |

9

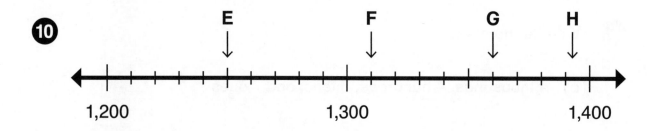

What numbers are indicated by each letter?

10

What numbers are indicated by the letters E, F, and G?

(a) What numbers are indicated by the letters E, F, and G?

(b) H is nearest to the tick mark for what number?

11 What are the greatest and least 4-digit numbers you can make using all the digits 4, 0, 2, and 8?

Exercise 6 • page 18

Lesson 7
Number Patterns

Think

I have 6,482 stickers.

(a) Sofia's mother gave her 1,000 more stickers.
How many stickers does Sofia have in all?

(b) Sofia then gave 100 stickers to her friend.
How many stickers does she have now?

Learn

(a)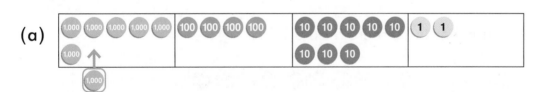

6,482 + **1**,000 = **7**,482

After her mother gives her 1,000 stickers, Sofia has _____ stickers.

_____

(b)

7,482 − 100 = 7,**3**82

After she gives 100 stickers to her friend, Sofia has _____ stickers.

Do

1 (a) What number is 1,000 more than 5,763?

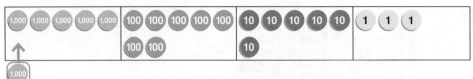

5 thousands + 1 thousand

$5{,}763 + 1{,}000 = \boxed{}$

(b) What number is 100 less than 5,763?

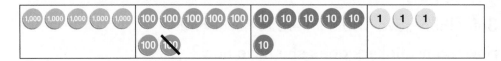

$5{,}763 - 100 = \boxed{}$

(c) What number is 10 more than 5,763?

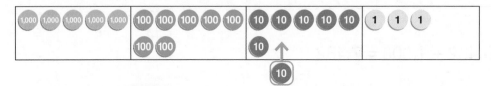

$5{,}763 + 10 = \boxed{}$

(d) What number is 1,000 less than 5,763?

(e) What number is 10 less than 5,763?

2 (a) What number is 100 more than 1,935?

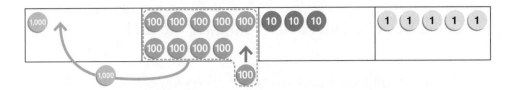

$1,935 + 100 =$ ⬚   19 hundreds + 1 hundred

(b) What number is 100 less than 1,935?

3 (a) What number is 100 less than 6,000?

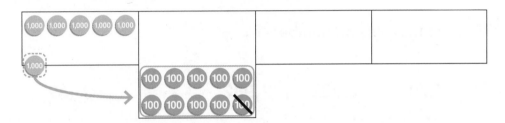

$6,000 - 100 =$ ⬚   60 hundreds − 1 hundred

(b) What number is 10 less than 6,000?

   600 tens − 1 ten

(c) What number is 1 less than 6,000?

4 (a) Count on by thousands from 2,740 to 7,740.

    (b) Count back by thousands from 5,281 to 281.

    (c) Count on by hundreds from 3,760 to 4,260.

    (d) Count back by hundreds from 9,207 to 8,707.

    (e) Count on by tens from 2,980 to 3,030.

    (f) Count back by tens from 6,025 to 5,985.

5 What are the missing numbers?

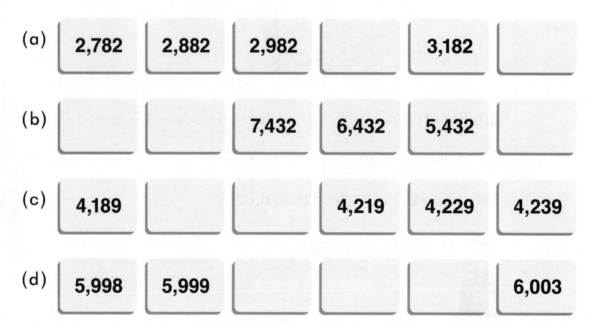

(a) | 2,782 | 2,882 | 2,982 | | 3,182 | |

(b) | | | 7,432 | 6,432 | 5,432 | |

(c) | 4,189 | | | 4,219 | 4,229 | 4,239 |

(d) | 5,998 | 5,999 | | | | 6,003 |

Exercise 7 • page 22

     1-7 Number Patterns

Think

There are 2,736 people watching a baseball game.
About how many people, to the nearest thousand, are watching the game?

Learn

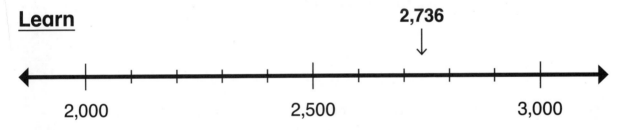

2,000   2,500   3,000

2,736 is between 2,000 and 3,000.
It is greater than 2,500 so it is nearer to...

2,736 is 3,000 when **rounded** to the nearest thousand.

When we say 2,736 is about 3,000,
we are rounding 2,736 to the nearest thousand.

Do

1 Round each number to the nearest thousand.

(a) 7,263

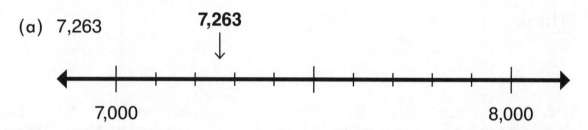

(b) 2,500

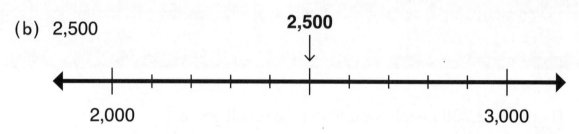

2,500 is halfway between 2,000 and 3,000.
We round to 3,000.

2 Locate about where each number is on the number line.
Round each number to the nearest thousand.

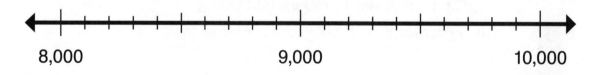

(a) 9,467   (b) 8,005   (c) 8,616   (d) 9,762

3 Round each number to the nearest thousand.

(a) 4,389   (b) 2,506   (c) 6,084   (d) 1,578

Exercise 8 • page 25

Think

(a) There are 2,736 people watching the baseball game.
About how many people, to the nearest hundred, are watching the game?

(b) If 14 more people come, how many will be watching the game?
Round this new number to the nearest hundred.

Learn

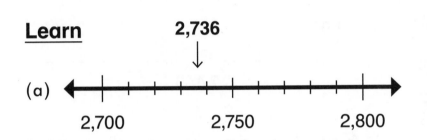

(a)

2,736 is between
2,700 and 2,800.
It is less than 2,750
so it is nearer to…

2,736 is 2,700 when rounded to the nearest hundred.

(b) 2,736 + 14 = 2,750

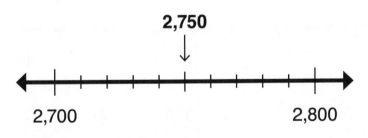

2,750 is halfway between 2,700 and 2,800.
I should round to…

2,750 is 2,800 when rounded to the nearest hundred.

Do

1 Round each number to the nearest hundred.

(a) 780

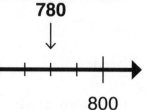

(b) 1,997

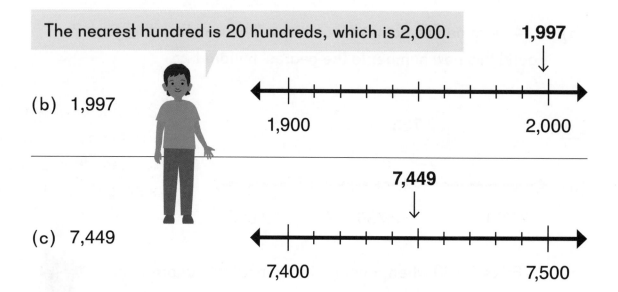

The nearest hundred is 20 hundreds, which is 2,000.

(c) 7,449

2 Locate about where each number is on the number line.
Round each number to the nearest hundred.

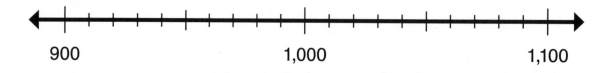

(a) 952   (b) 925   (c) 1,070   (d) 1,007

3 Round each number to the nearest hundred.

(a) 250   (b) 710   (c) 493   (d) 629

(e) 6,353   (f) 2,450   (g) 6,972   (h) 9,035

Exercise 9 • page 27

Think

(a) There are 2,736 people watching the baseball game.
About how many people, to the nearest ten, are watching the game?

(b) 1,004 of the people are children.
What is 1,004 rounded to the nearest ten?

Learn

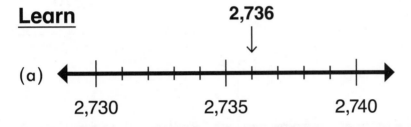

(a)

2,736 is between 2,730 and 2,740.
It is greater than 2,735 so it is nearer to...

2,736 is 2,740 when rounded to the nearest ten.

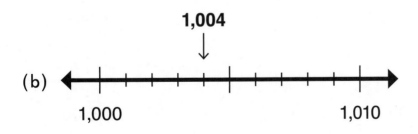

(b)

1,004 is 1,000 when rounded to the nearest ten.

Do

1 Round each number to the nearest ten.

(a) 38

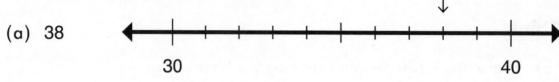

(b) 563

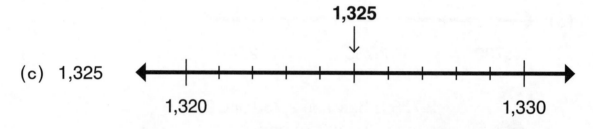

(c) 1,325

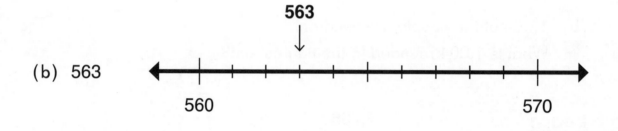

 1,325 is halfway between 1,320 and 1,330. What number should we round to?

(d) 4,995

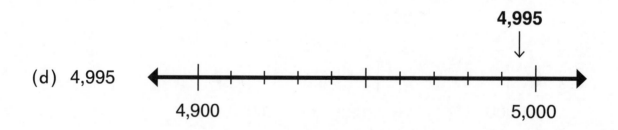

2 Locate about where each number is on the number line.
Round each number to the nearest ten.

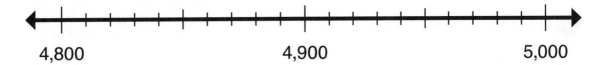

4,800　　　　　　　　4,900　　　　　　　　5,000

(a) 4,805　　　　　　　(b) 4,996

(c) 4,950　　　　　　　(d) 4,872

3 Round each number to the nearest ten.

(a) 48　　　　　　(b) 75　　　　　　(c) 96

(d) 23　　　　　　(e) 231　　　　　(f) 897

(g) 495　　　　　(h) 903　　　　　(i) 4,166

(j) 1,585　　　　(k) 7,097　　　　(l) 3,996

4 Eli the elephant weighs 6,527 pounds.
Round his weight to the nearest thousand,
the nearest hundred,
and the nearest ten.

Exercise 10 • page 30

1 Find the number that is...

(a) 1,000 more than 6,307.

(b) 1,000 less than 3,008.

(c) 100 more than 5,925.

(d) 100 less than 7,777.

(e) 10 more than 8,943.

(f) 10 less than 2,301.

2 (a) 5,884 − 1,000 = ▢   (b) 6,692 + 100 = ▢

(c) 4,044 + 10 = ▢   (d) 1,068 − 100 = ▢

(e) 2,538 − ▢ = 2,438   (f) 1,997 + ▢ = 2,007

(g) 8,787 + ▢ = 9,787   (h) 5,000 − ▢ = 4,999

3 What are the missing numbers?

(a)

3,334		3,134			2,834

(b)

1,435	1,335	1,235		1,035	

(c)

2,020			1,990	1,980	

(d)

	4,999		5,001		5,003

(e)

	5,285	5,295		5,315	

4 (a) Count on by thousands from 2,910 to 7,910.

(b) Count back by thousands from 5,371 to 371.

(c) Count on by hundreds from 4,780 to 5,280.

(d) Count back by hundreds from 8,105 to 7,605.

5 Locate 7,365 on each number line.
Then round 7,365 to the nearest...

(a) Thousand

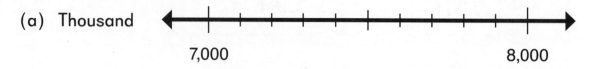

7,000   8,000

(b) Hundred

7,300   7,400

(c) Ten

7,360   7,370

6 Round each number to the nearest thousand.

(a) 4,209   (b) 2,070   (c) 3,505   (d) 900

7 Round each number to the nearest hundred.

(a) 670   (b) 3,250   (c) 9,084   (d) 5,555

8 Round each number to the nearest ten.

(a) 163   (b) 3,287   (c) 8,005   (d) 4,996

Exercise 11 • page 33

Chapter 2

Addition and Subtraction — Part 1

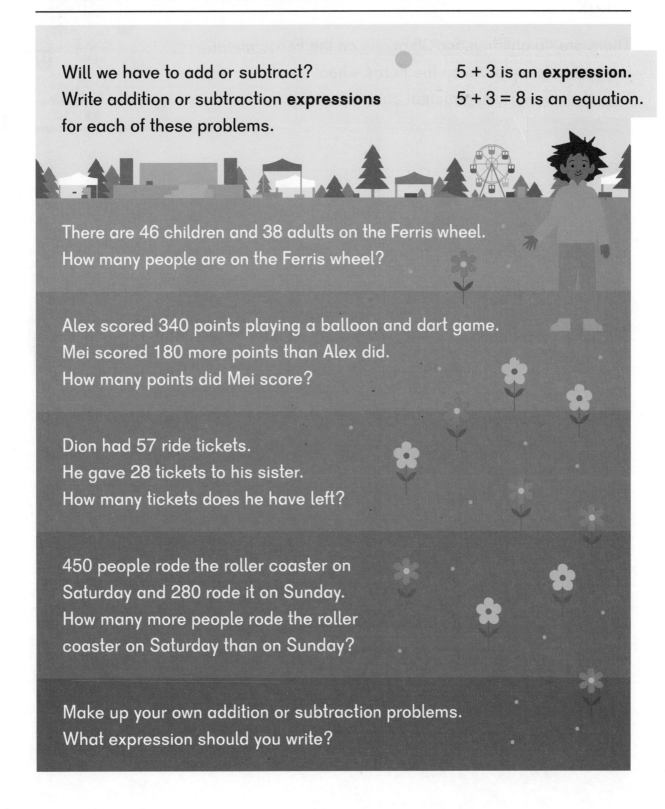

Will we have to add or subtract?
Write addition or subtraction **expressions** for each of these problems.

5 + 3 is an **expression**.
5 + 3 = 8 is an equation.

There are 46 children and 38 adults on the Ferris wheel.
How many people are on the Ferris wheel?

Alex scored 340 points playing a balloon and dart game.
Mei scored 180 more points than Alex did.
How many points did Mei score?

Dion had 57 ride tickets.
He gave 28 tickets to his sister.
How many tickets does he have left?

450 people rode the roller coaster on Saturday and 280 rode it on Sunday.
How many more people rode the roller coaster on Saturday than on Sunday?

Make up your own addition or subtraction problems.
What expression should you write?

Think

There are 46 children and 38 adults on the Ferris wheel.
How many people are on the Ferris wheel?
Find the answer using mental calculation.

Learn

46 + 38

30   8

46 + 30 = 76
76 + 8 = 84

46 + 38

44   2

38 + 2 = 40
44 + 40 = 84

40 is **2** more than **38**.
46 + 40 = 86
86 − 2 = 84

46 $\xrightarrow{+40}$ 86 $\xrightarrow{-2}$ 84

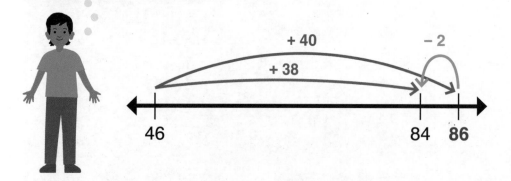

46 + 38 = ⬜ | ⬜ people are on the Ferris wheel.

Do

1 (a) 57 + 32 = ▢

$$57 \xrightarrow{\;+30\;} 87 \xrightarrow{\;+2\;} ?$$

(b) 35 + 56 = ▢

$$35 \xrightarrow{\;+50\;} ? \xrightarrow{\;+6\;} ?$$

(c) 37 + 36 = ▢

37 + 36
／＼
3   33

(d) 74 + 19 = ▢

$$74 \xrightarrow{\;+20\;} 94 \xrightarrow{\;-1\;} ?$$

I don't have to regroup!

2 Use mental calculation to find the value.

(a) 57 + 8   (b) 49 + 30   (c) 76 + 22

(d) 42 + 18   (e) 53 + 29   (f) 36 + 45

(g) 76 + 24   (h) 13 + 69   (i) 28 + 25

Exercise 1 • page 37

Think

Alex scored 340 points playing a balloon and dart game.
Mei scored 180 more points than Alex did.
How many points did Mei score?
Find the answer using mental calculation.

Learn

340 + 180

100   80

340 + 100 = 440
440 + 80 = 520

340 + 180

320   20

20 + 180 = 200
320 + 200 = 520

200 is **20** more than **180**.
340 + 200 = 540
540 − 20 = 520

340 $\xrightarrow{+200}$ 540 $\xrightarrow{-20}$ 520

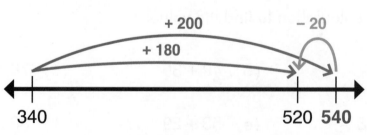

+ 200
+ 180
− 20

340   520 540

340 + 180 = ▢ | Mei scored ▢ points.

Do

1 480 + 340 = [ ]

+300 +40
480 ⟶ ? ⟶ ?

2 (a) 770 + 60 = [ ]   (b) 170 + 680 = [ ]

770 + 60
/ \
30 30

170 + 680
/ \
30 650

3 (a) 570 + 290 = [ ]   (b) 450 + 70 = [ ]

+300 −10
570 ⟶ ? ⟶ ?

+100 −30
450 ⟶ ? ⟶ ?

4 Use mental calculation to find the value.

(a) 220 + 70   (b) 450 + 50   (c) 80 + 170

(d) 310 + 260   (e) 640 + 160   (f) 390 + 540

(g) 270 + 480   (h) 830 + 170   (i) 180 + 760

Exercise 2 • page 41

Think

Dion had 57 ride tickets.
He gave 28 tickets to his sister.
How many tickets does he have left?
Find the answer using mental calculation.

Learn

$57 - 28$
↙ ↘
20   8

$57 - 20 = 37$
$37 - 8 = 29$

$57 - 28$
↙ ↘
27   30

$30 - 28 = 2$
$27 + 2 = 29$

30 is **2** more than **28**.
$57 - 30 = 27$
$27 + 2 = 29$

$$57 \xrightarrow{-30} 27 \xrightarrow{+2} 29$$

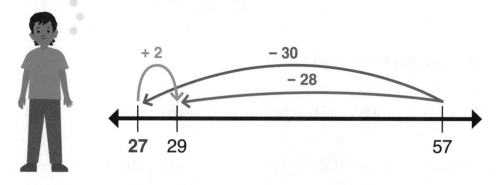

$57 - 28 = \boxed{}$ | He has $\boxed{}$ tickets left.

Do

1 (a) 76 − 23 =

$$76 \xrightarrow{-20} 56 \xrightarrow{-3} ?$$

(b) 62 − 47 =

$$62 \xrightarrow{-40} ? \xrightarrow{-7} ?$$

2 81 − 56 =

81 − 56
⟋ ⟍
21   60

3 74 − 19 =

$$74 \xrightarrow{-20} 54 \xrightarrow{+1} ?$$

I don't have to regroup!

4 Use mental calculation to find the value.

(a) 87 − 34   (b) 97 − 67   (c) 90 − 18

(d) 70 − 47   (e) 61 − 26   (f) 73 − 39

(g) 84 − 57   (h) 90 − 79   (i) 53 − 28

Exercise 3 • page 43

Think

450 people rode on the roller coaster on Saturday and 280 rode on it on Sunday.
How many more people rode the roller coaster on Saturday than on Sunday?
Find the answer using mental calculation.

Learn

450 − 280
／　＼
200　80

450 − 200 = 250
250 − 80 = 170

450 − 280
／　＼
150　300

300 − 280 = 20
150 + 20 = 170

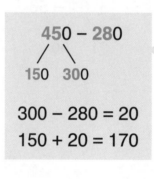

300 is **20** more than **280**.
450 − 300 = **150**
150 + **20** = 170

$$450 \xrightarrow{-300} 150 \xrightarrow{+20} 170$$

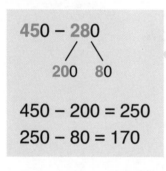

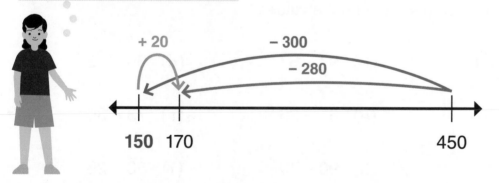

+ 20

− 300
− 280

150　170　　　　　450

450 − 280 = ☐ | ☐ more people rode the roller coaster on Saturday.

Do

1   790 − 250 = ▢

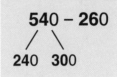

$$790 \xrightarrow{-200} ? \xrightarrow{-50} ?$$

2   (a)   120 − 50 = ▢   (b)   540 − 260 = ▢

120 − 50
20   30

540 − 260
240   300

3   (a)   510 − 90 = ▢   (b)   820 − 280 = ▢

$$510 \xrightarrow{-100} ? \xrightarrow{+10} ?$$

$$820 \xrightarrow{-300} ? \xrightarrow{+20} ?$$

4   Use mental calculation to find the value.

(a)   370 − 50   (b)   540 − 70   (c)   700 − 80

(d)   890 − 320   (e)   510 − 160   (f)   830 − 490

(g)   600 − 350   (h)   540 − 260   (i)   410 − 370

Exercise 4 • page 47

Think

Find pairs of numbers that make 100.

57	15	64	79	62

38	21	43	36	85

What are the totals of the tens and the ones
of the pairs of numbers that make 100?

Learn

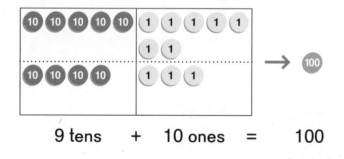

9 tens + 10 ones = 100

$50 + 40 =$

$7 + 3 =$

$57 + 43 = 100$

57   43

9 tens   10 ones

5 tens + ? tens = 9 tens
7 ones + ? ones = 10 ones

Do

1 (a) 37 + ▢ = 100

(b) 100 − 37 = ▢

(c) 200 − 37 = ▢

(d) 800 − 37 = ▢

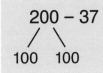

3 tens + ? tens = 9 tens
7 ones + ? ones = 10 ones

200 − 37
⁄ \
100   100

2 640 + 360 = ▢

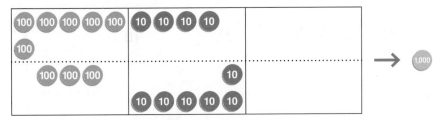

9 hundreds +   10 tens   =   1,000

600 + 300 = ▢

40 + 60 = ▢

640 + 360 = 1,000

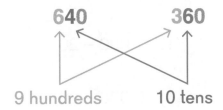

640   **360**

9 hundreds   10 tens

6 hundreds + 3 hundreds = ? hundreds
4 tens + 6 tens = ? tens

3 (a) 260 + [ ] = 1,000

(b) 1,000 − 260 = [ ]

(c) 5,000 − 260 = [ ]

2 hundreds + ? hundreds = 9 hundreds
6 tens + ? tens = 10 tens

5,000 − 260

4,000   1,000

4 Find pairs of numbers that make 1,000.

| 530 | 140 | 290 | 650 | 480 |

| 350 | 520 | 860 | 470 | 710 |

5 Use mental calculation to find the value.

(a) 100 − 79

(b) 100 − 47

(c) 100 − 16

(d) 500 − 24

(e) 300 − 8

(f) 900 − 52

(g) 1,000 − 700

(h) 1,000 − 360

(i) 1,000 − 20

(j) 6,000 − 240

(k) 4,000 − 590

(l) 6,000 − 80

Exercise 5 • page 49

Think

125 children and 98 adults ate snow cones.

(a) How many people ate snow cones?

(b) How many more children than adults ate snow cones?

Learn

(a)

$$125 + 98$$

123   2

$2 + 98 = 100$
$123 + 100 = 223$

$125 + 98 = \boxed{}$

---

100 is **2** more than **98**.
$125 + 100 = 225$
$225 - 2 = 223$

$\boxed{}$ people ate snow cones.

(b)

$$125 - 98$$

25   100

$100 - 98 = 2$
$25 + 2 = 27$

$125 - 98 = \boxed{}$

---

100 is **2** more than **98**.
$125 - 100 = 25$
$25 + 2 = 27$

$\boxed{}$ more children than adults ate snow cones.

Do

1 (a) $526 + 97 =$ ▢

(b) $526 + 397 =$ ▢

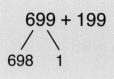
$$526 \xrightarrow{+400} ? \xrightarrow{-3} ?$$

2 (a) $526 - 97 =$ ▢

(b) $526 - 397 =$ ▢

$$526 \xrightarrow{-400} ? \xrightarrow{+3} ?$$

3 (a) $498 + 54 =$ ▢   (b) $699 + 199 =$ ▢

$$498 + 54$$
$$\diagup\ \diagdown$$
$$2 \quad 52$$

$$699 + 199$$
$$\diagup\ \diagdown$$
$$698 \quad 1$$

4 Use mental calculation to find the value.

(a) $99 + 98$ (b) $478 + 98$ (c) $155 + 497$

(d) $104 - 97$ (e) $510 - 96$ (f) $830 - 499$

(g) $463 + 98$ (h) $207 - 99$ (i) $326 + 97$

(j) $387 - 196$ (k) $296 + 407$ (l) $984 - 598$

Exercise 6 · page 51

2-6 Strategies for Numbers Close to Hundreds

1 Use mental calculation to find the value.

(a) 38 + 47 (b) 83 − 25 (c) 470 + 60

(d) 740 − 90 (e) 150 + 680 (f) 740 − 260

(g) 400 − 32 (h) 1,000 − 480 (i) 330 + 390

(j) 464 + 96 (k) 498 + 307 (l) 584 − 98

(m) 892 − 297 (n) 600 − 162 (o) 5,000 − 97

2 Write an expression for each of the following.
Use mental calculation to find the value.

(a) An unlimited ride wristband for 1 day costs $39.
A sheet of 80 ride tickets costs $75.
How much more is the sheet of tickets than the wristband?

(b) A man sold 450 balloons in the morning and 390 in the afternoon.
How many balloons did he sell that day?

(c) 195 people rode on the pirate ship in the morning.
97 rode on it in the afternoon.
How many people rode on the pirate ship that day?

Exercise 7 • page 53

Lesson 8
Sum and Difference

Think

Cut paper strips to show 231 and 563.
Use the paper strips to model the following:

(a) Find the sum of 231 and 563.

(b) Find the difference between 231 and 563.

Draw the models and label each part.
Use a question mark for what needs to be found.
Then find the value represented by the question mark.

Which number is represented by the longer strip?

Learn

231 + 563 =

The **sum** of 231 and 563 is 794.

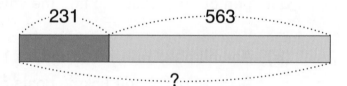

563 − 231 =

The **difference** between 231 and 563 is 332.

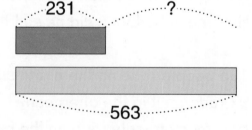

Do

1

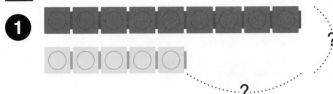

?   9 + 5 =

The sum of 9 and 5 is   .

9 − 5 =

The difference between 9 and 5 is   .

2 Find the sum of 75 and 47.

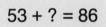
Part + Part = Whole

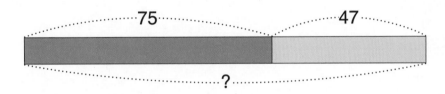

75 + 47 =

3 The sum of two numbers is 86.
One number is 53.
Find the other number.

53 + ? = 86

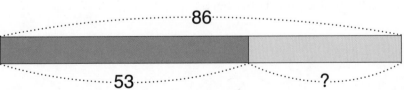

86 − 53 =   Whole − Part = Part

4 Find the difference between 75 and 47.

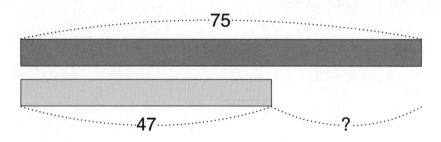

$75 - 47 =$

5 The difference between two numbers is 80.
The greater number is 120.

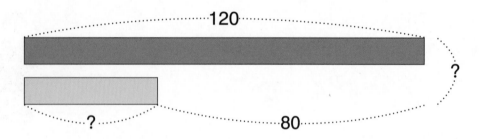

(a) What is the lesser number?

$120 - 80 =$

$120 - ? = 80$

(b) What is the sum of the two numbers?

$120 +$ $=$

6 The sum of two numbers is 100.
One number is 35.

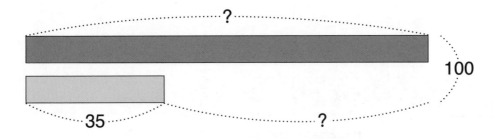

(a) What is the other number?

$$100 - 35 = \boxed{}$$

(b) What is the difference between the two numbers?

$$\boxed{} - 35 = \boxed{}$$

7 The difference between two numbers is 43.
The lesser number is 98.

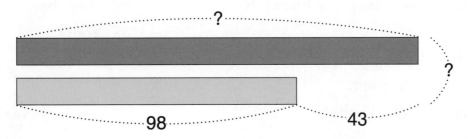

(a) What is the greater number?   ? − 98 = 43

$$98 + 43 = \boxed{}$$

(b) What is the sum of the two numbers?

Exercise 8 • page 55

Think

(a) A vendor had 150 candles.
She sold 98 candles.
How many candles does she have left?

(b) A vendor had 150 candles.
She has 52 candles left.
How many candles did she sell?

(c) A vendor sold 98 candles.
She has 52 candles left.
How many candles did she have at first?

Draw a bar model and write an expression for each problem.
Solve each problem.

How are the
problems similar?
How are they
different?

<u>Learn</u>

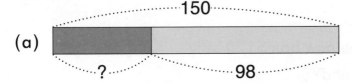

(a)

$150 - 98 = $ ⬚   Whole − Part = Part

She has ⬚ candles left.

(b)

$150 - 52 = $ ⬚   Whole − Part = Part

She sold ⬚ candles.

(c)

$98 + 52 = $ ⬚   Part + Part = Whole

She had ⬚ candles at first.

We can use a part-whole
model for all three problems.

Do

1 Abigail had $450.

After spending some of it, she now has $170 left.

How much money did she spend?

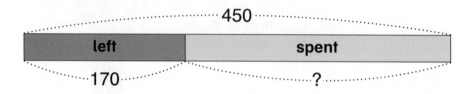

$450 - 170 = \boxed{}$

She spent $\$\boxed{}$.

2

Sasha collects foreign coins.

She has 90 coins from the southern part of Africa.

32 of them are from Botswana and 18 are from Zimbabwe.

The rest of the coins are from South Africa.

(a) How many of the coins are not from South Africa?

(b) How many of the coins are from South Africa?

3 Draw models and solve.

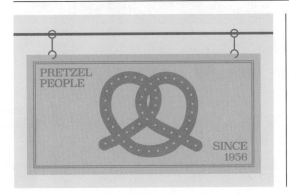

(a) There are 150 girls and 160 boys in the school bands.
How many students are in the bands?

(b) There are 640 people at the school concert.
180 of them are adults and the rest are children.
How many children are at the concert?

(c) A bakery had 650 pretzels to sell at the start of the day.
There were 20 pretzels left at the end of the day.
How many pretzels were sold that day?

(d) A book has 600 pages.
Jack read 430 pages.
How many pages does he still have to read?

Exercise 9 • page 60

Think

(a)   360 plain funnel cakes were sold at a booth at the fair.
      550 more berry funnel cakes were sold than plain funnel cakes.
      How many berry funnel cakes were sold?

(b)   910 berry funnel cakes were sold at a booth at the fair.
      550 fewer plain funnel cakes were sold than berry funnel cakes.
      How many plain funnel cakes were sold?

(c)   910 berry funnel cakes were sold at a booth at the fair.
      360 plain funnel cakes were sold.
      How many more berry funnel cakes were sold than plain funnel cakes?

Draw a bar model and write an expression for each problem.
Solve each problem.

Learn

(a) **plain**   360   550

berry   ?

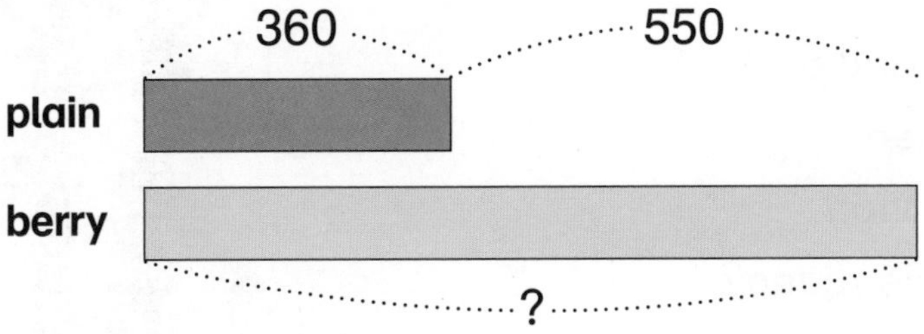

360 + 550 = ⬜

⬜ berry funnel cakes were sold.

(b) **plain**   ?   550

berry   910

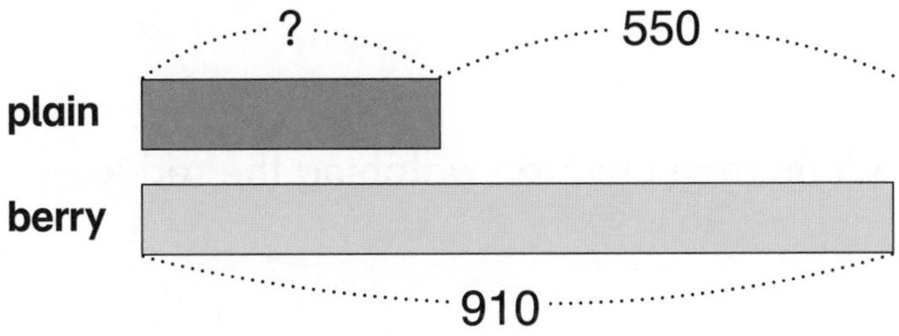

910 − 550 = ⬜

⬜ plain funnel cakes were sold.

(c) **plain**   360   ?

berry   910

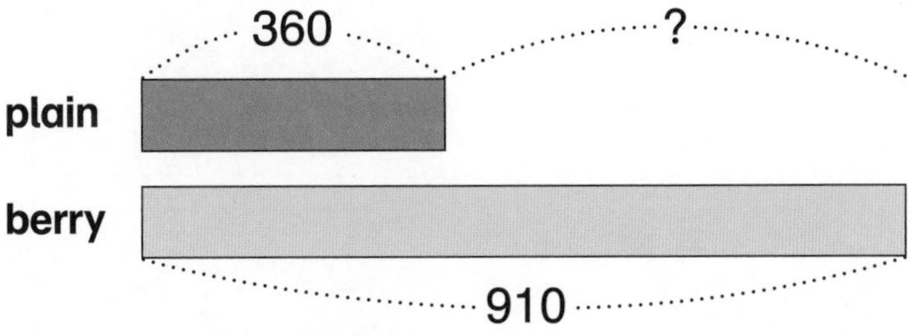

910 − 360 = ⬜

We can use a comparison model for all three problems.

⬜ more berry funnel cakes were sold than plain funnel cakes.

Do

1 There are 350 children and 820
adults watching the rodeo.
How many more adults than
children are watching the rodeo?

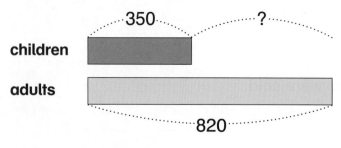

820 – 350 = [ ]

There are [ ] more adults than children watching the rodeo.

2 A flower shop has 680 dahlias.
It has 290 more dahlias than tulips.
How many tulips does it have?

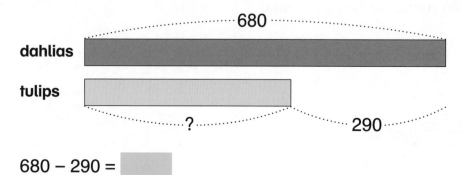

680 – 290 = [ ]

It has [ ] tulips.

3 Draw models and solve.

(a) The flower shop has 430 roses and 270 carnations.
How many more roses does it have than carnations?

(b) The shop has 320 asters.
It has 197 more daisies than asters.
How many daisies does it have?

(c) The shop has 120 peonies.
It has 70 more peonies than lilies.
How many lilies does it have?

(d) A Joyful bouquet costs $71.
A Sunshine bouquet costs $49.
How much less does a Sunshine bouquet cost than a Joyful bouquet?

Exercise 10 • page 62

Think

A booth sold 540 corn dogs in the afternoon.

It sold 190 more corn dogs in the afternoon than in the morning.

How many corn dogs were sold that day?

Learn

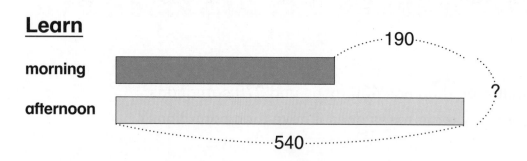

I need to find the number sold in the morning first.

Number sold in the morning

540 − 190 = 350

Total sold

350 + 540 =

There were _____ corn dogs sold that day.

Do

1 There are 730 students at Washington School.

There are 80 fewer students at Washington School than at Lincoln School.

How many students are there at the two schools altogether?

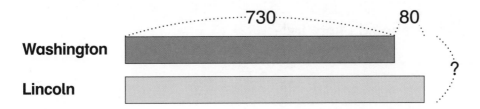

Washington

Lincoln

730 + 80 = ⬜

⬜ + 730 = ⬜

There are ⬜ students at the two schools.

Find the number at Lincoln School first.

2 There are 69 students in the three third grade classes.

24 students are in 3A and 22 students in 3C.

How many students are in 3B?

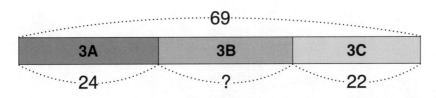

3A	3B	3C

69 − 24 = 45
45 − 22 = ?

24 + 22 = 46
69 − 46 = ?

3 There are 810 students at Deer Harbor Academy.
360 of them are in the Upper School.
The rest are in the Lower School.
How many more students are in the Lower School than the Upper School?

What should I find first?

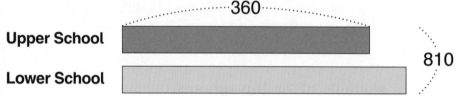

4 Chapa saved $32.
Emily saved $14 more than Chapa.
Ivy saved $17 less than Emily.

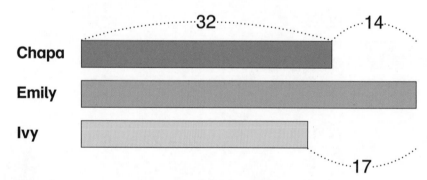

(a) How much money did Ivy save?

(b) How much money did the three girls save altogether?

5 Draw models and solve.

(a) 97 students are in grade 6 at Washington School.
30 more students are in grade 5 than grade 6.
How many students are in both grades?

(b) The students in a cooking club baked 250 zucchini muffins.
They baked 80 fewer pumpkin muffins than zucchini muffins.
How many muffins did they bake in all?

(c) The school made $530 at a bake sale.
$310 came from selling pumpkin muffins.
The rest came from selling zucchini muffins.
How much less was made from selling zucchini
muffins than pumpkin muffins?

(d) Maya sold 14 muffins at the bake sale.
Adam sold 24 more muffins than Maya.
Ella sold 15 fewer muffins than Adam.
How many muffins did Ella sell?

Exercise 11 • page 64

1 Cooper has 350 raffle tickets to sell.
He sold 120 tickets yesterday and 140 tickets today.
How many raffle tickets has he not sold?

2 Bayla spent $470 on a game console.
She spent $280 less on a controller than on the game console.
How much did she spend in all?

3 A guitar costs $50 more than a banjo.
The banjo costs $140.
How much do both instruments cost altogether?

4 The difference between two numbers is 90.
The greater number is 310.
What is the sum of the two numbers?

5 The difference between two numbers is 30.
The lesser number is 90.
What is the sum of the two numbers?

6 The sum of two numbers is 110.
One of the numbers is 50.
What is the difference between the two numbers?

Chapter 3

Addition and Subtraction — Part 2

$349

$3,685

$294

$100

$639

$849

$1,189

$2,899

DM

Pick two instruments. Do you know how to find the sum of their costs and the difference between their costs?

Which numbers do I already know how to add or subtract?

Round the numbers, then add or subtract to find the approximate sum and difference.

Which place should I round each number to?

Lesson 1
Addition with Regrouping

Think

A trumpet costs $3,685.
A piano costs $2,947 more than the trumpet.
How much does the piano cost?

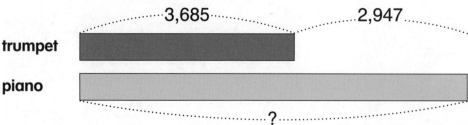

3,685   2,947

trumpet

piano

?

There are more than 10
hundreds, tens, and ones.

Learn

3,685 + 2,947

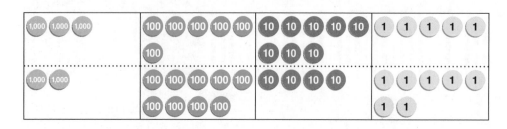

Add the ones and regroup.

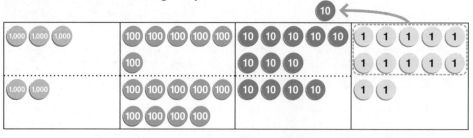

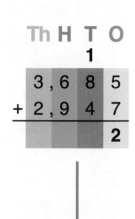

Th **H** **T** **O**
 1

  3 , 6 8 5
+ 2 , 9 4 7
                2

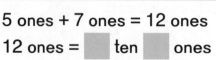

5 ones + 7 ones = 12 ones

12 ones = ⬜ ten ⬜ ones

Add the tens and regroup.

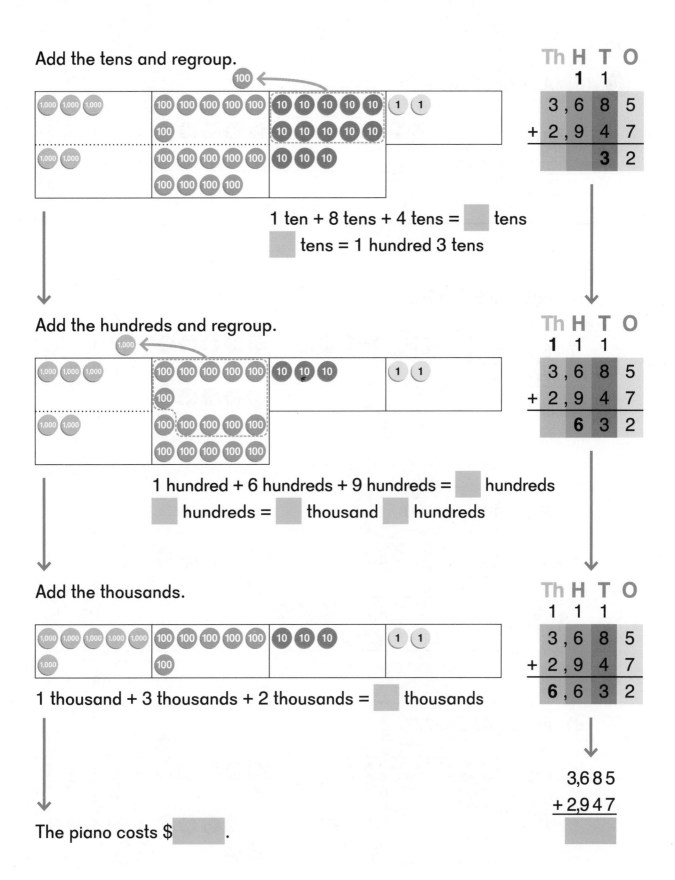

1 ten + 8 tens + 4 tens = ☐ tens

☐ tens = 1 hundred 3 tens

Th	H	T	O
	1	1	
3,	6	8	5
+ 2,	9	4	7
		3	2

Add the hundreds and regroup.

1 hundred + 6 hundreds + 9 hundreds = ☐ hundreds

☐ hundreds = ☐ thousand ☐ hundreds

Th	H	T	O
1	1	1	
3,	6	8	5
+ 2,	9	4	7
	6	**3**	2

Add the thousands.

1 thousand + 3 thousands + 2 thousands = ☐ thousands

Th	H	T	O
1	1	1	
3,	6	8	5
+ 2,	9	4	7
6,	**6**	**3**	2

The piano costs $ ☐ .

```
  3,685
+ 2,947
  ☐
```

Do

  (a) Add 7,468 and 7.

$$\begin{array}{r} 7{,}468 \\ +\quad 7 \\ \hline \end{array}$$

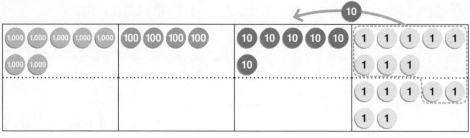

(b) Add 7,468 and 70.

$$\begin{array}{r} 7{,}468 \\ +\quad 70 \\ \hline \end{array}$$

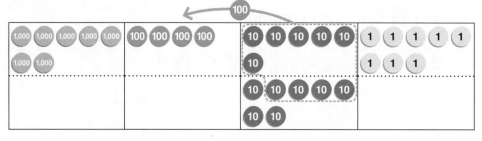

(c) Add 7,468 and 700.

$$\begin{array}{r} 7{,}468 \\ +\quad 700 \\ \hline \end{array}$$

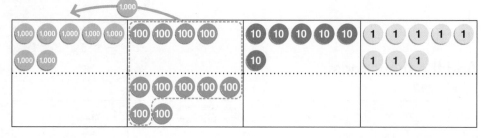

(d) Add 7,468 and 777.

$$\begin{array}{r} 7{,}468 \\ +\quad 777 \\ \hline \end{array}$$

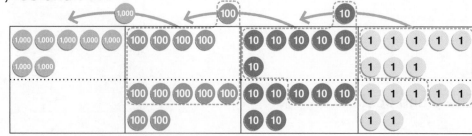

2 Find the value.

(a) 2,356 + 86

(b) 6,962 + 93

(c) 8,480 + 694

(d) 7,028 + 909 + 60

(e) 5,272 + 3,764

(f) 3,547 + 1,793 + 436

3 What number is 888 more than 6,666?

4 What are the missing digits?

(a)
```
    ▢ , 4  2  ▢
  +    7  ▢  8
    5 , ▢  5  1
```

(b)
```
    6 , ▢  9  5
  + 2 , 6  0  ▢
    ▢ , 0  ▢  0
```

5 A school bought new musical equipment for the jazz band.
The amplifier cost $1,189, the conga drum cost $639,
and the chimes cost $294.
How much did all three instruments cost?

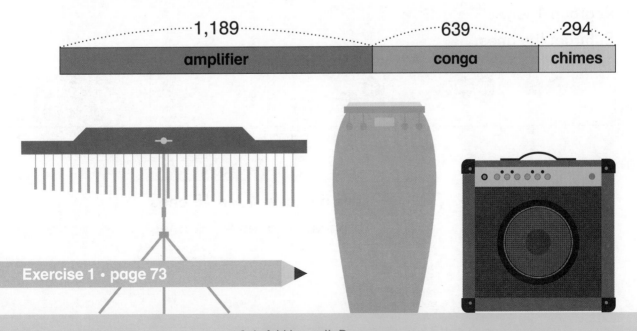

Exercise 1 • page 73

Think

There are 4,325 people watching a jazz band performance.
2,468 are adults and the rest are children.
How many children are there?

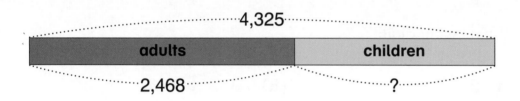

Learn

4,325 − 2,468

I will need more ones, tens, and hundreds.

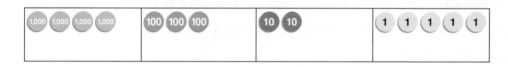

Regroup 1 ten.
Subtract the ones.

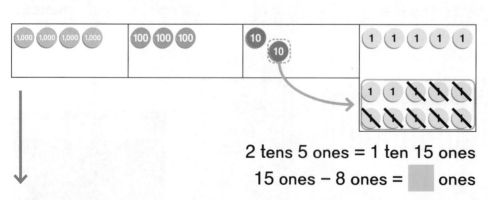

2 tens 5 ones = 1 ten 15 ones

15 ones − 8 ones = ☐ ones

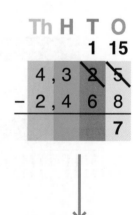

Regroup 1 hundred.

Subtract the tens.

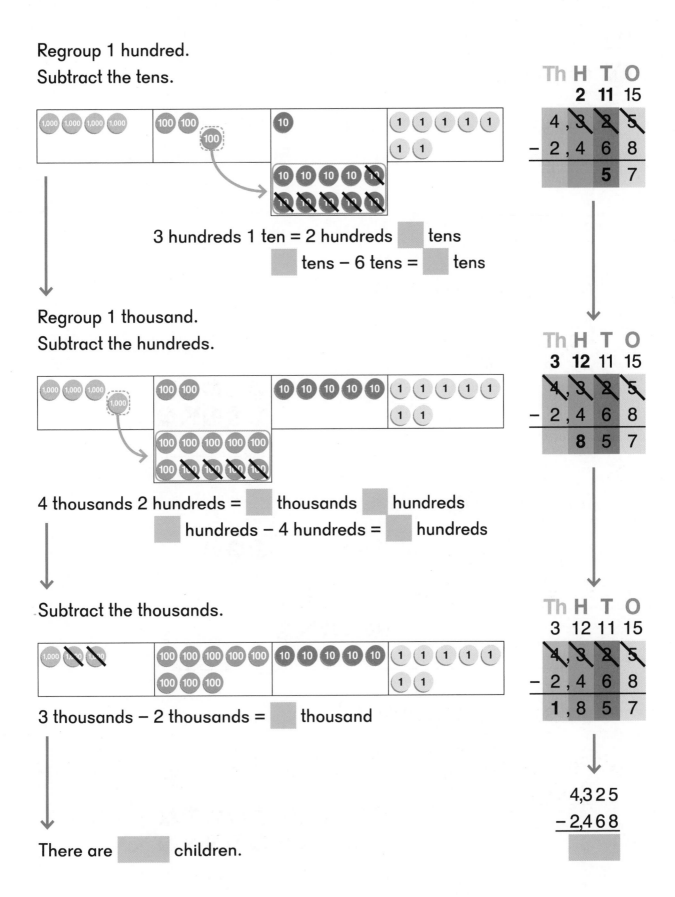

3 hundreds 1 ten = 2 hundreds ▢ tens

▢ tens − 6 tens = ▢ tens

Regroup 1 thousand.

Subtract the hundreds.

4 thousands 2 hundreds = ▢ thousands ▢ hundreds

▢ hundreds − 4 hundreds = ▢ hundreds

Subtract the thousands.

3 thousands − 2 thousands = ▢ thousand

There are ▢ children.

$$\begin{array}{r} 4{,}325 \\ -\,2{,}468 \\ \hline \phantom{\rule{2em}{1em}} \end{array}$$

Do

1 (a) Subtract 8 from 9,235.

$$\begin{array}{r} 9,235 \\ -\quad\;\; 8 \\ \hline \end{array}$$

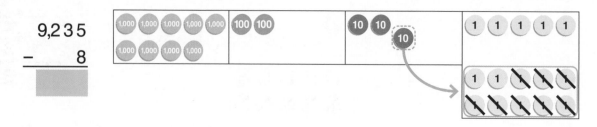

(b) Subtract 80 from 9,235.

$$\begin{array}{r} 9,235 \\ -\quad\; 80 \\ \hline \end{array}$$

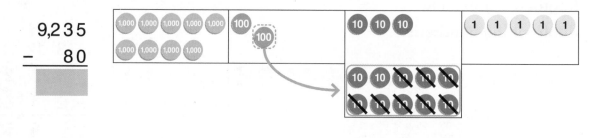

(c) Subtract 800 from 9,235.

$$\begin{array}{r} 9,235 \\ -\;\; 800 \\ \hline \end{array}$$

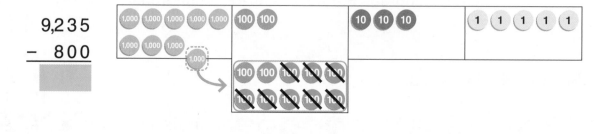

(d) Subtract 888 from 9,235.

$$\begin{array}{r} 9,235 \\ -\;\; 888 \\ \hline \end{array}$$

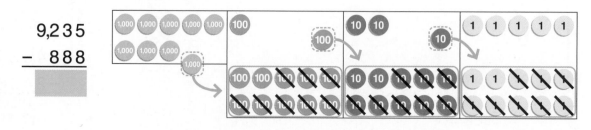

2 Subtract 2,683 from 6,148.

$$6,148$$
$$-2,683$$

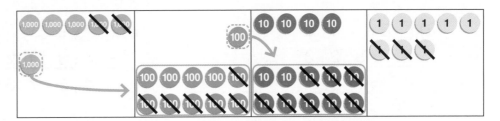

$$6,148$$
$$-2,683$$

$\times$

$$+2,683$$

Add to check the answer.

3 Find the value of 8,627 − 3,719.

$$8,627$$
$$-3,719$$

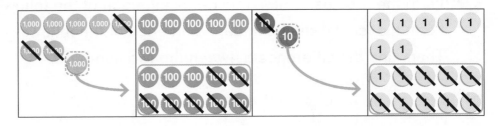

4 Find the value.

(a) 2,347 − 168

(b) 1,419 − 729

(c) 4,159 − 1,361

(d) 8,260 − 3,479

(e) 4,107 − 1,046

(f) 8,096 − 3,575

5 What are the missing digits?

(a)
```
   8 , ▨  4  9
 −    8  2  3
  ▨ , 7  2  ▨
```

(b)
```
   ▨ , 9  3  ▨
 − 3 , 3  ▨  2
   4 , 5  9  3
```

6 What are the missing numbers?

(a) 5,530 − ▨ = 168

(b) 5,092 − ▨ = 2,163

7 Form the greatest 4-digit number using each of the following digits once.
Form the least 4-digit number using each of the following digits once.
Then, find the difference between the two numbers.

| 4 | 8 | 1 | 3 |

8 The difference between two numbers is 1,498.
The greater number is 2,185.

(a) What is the other number?

(b) What is the sum of the
two numbers?

Exercise 2 • page 76

3-2 Subtraction with Regrouping — Part 1

Think

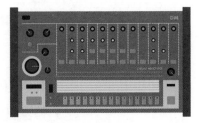

A keyboard and a drum machine together cost $5,003.

The keyboard cost $2,899.

How much does the drum machine cost?

Learn

5,003 – 2,899

1,000 1,000 1,000 1,000 1,000			1 1 1

> I need more ones, but there are no tens or hundreds.
> I have to regroup from the thousands.

Regroup 1 thousand.
Regroup 1 hundred.
Regroup 1 ten.

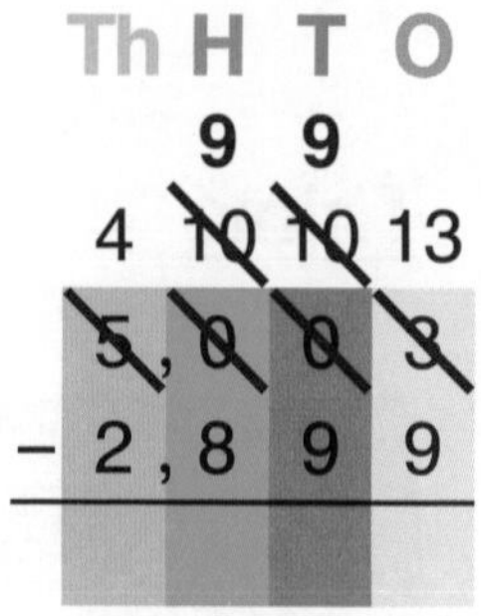

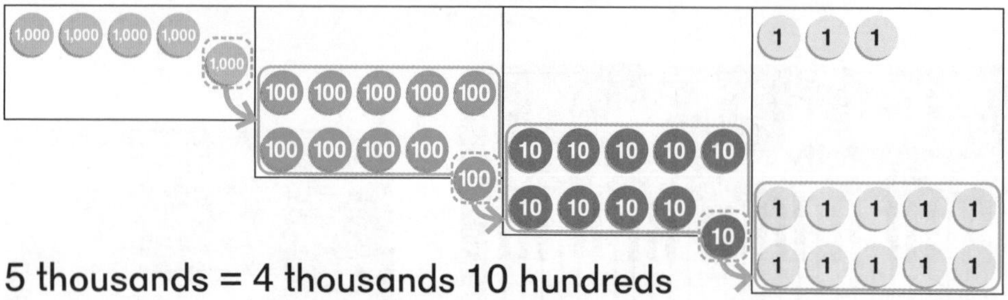

5 thousands = 4 thousands 10 hundreds

10 hundreds = 9 hundreds 10 tens

10 tens 3 ones = 9 tens ▢ ones

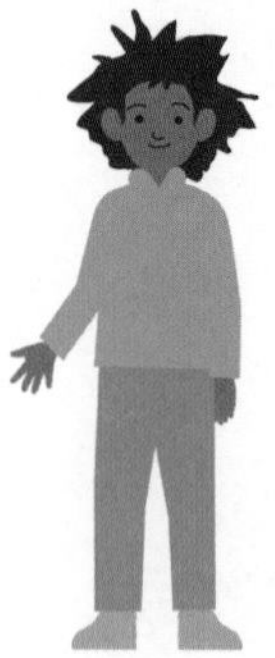

Now I can subtract the ones, the tens, and the hundreds.

Subtract the ones, the tens, the hundreds, and the thousands.

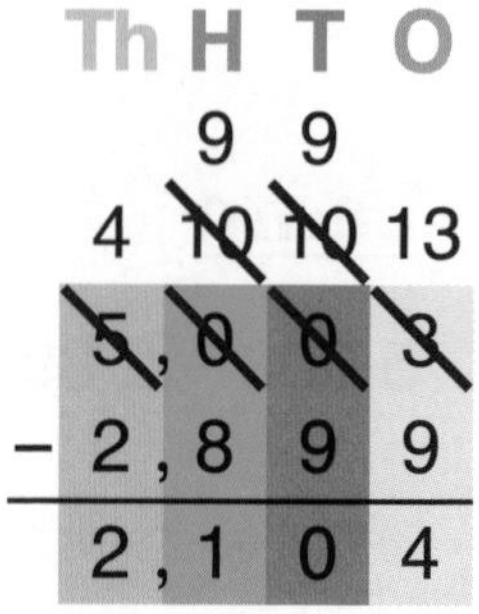

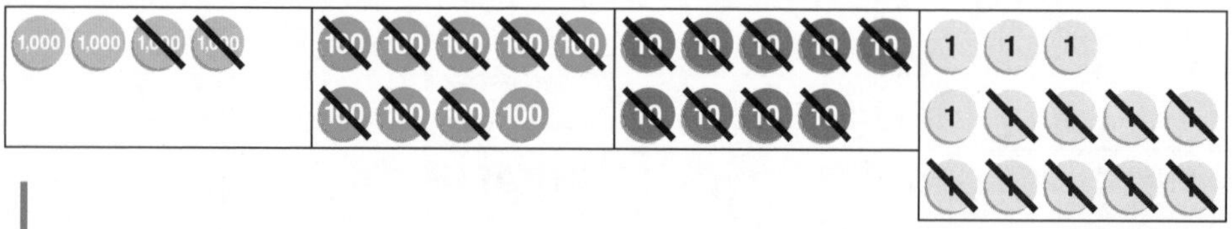

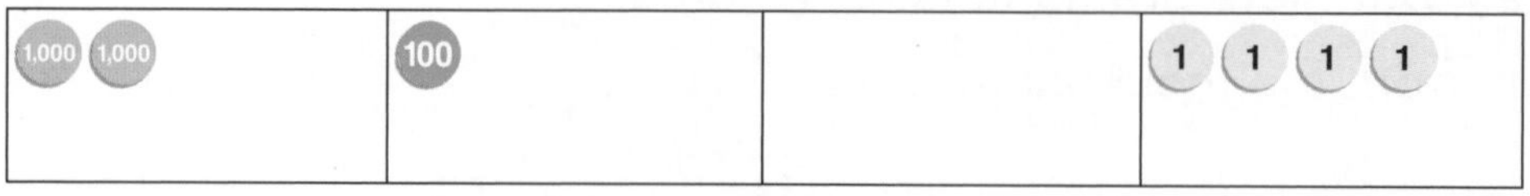

The drum machine cost $▢.

5,003
− 2,899
▢

Do

1 Subtract 1,458 from 5,036.

$$
\begin{array}{r}
5{,}0\,3\,6 \\
-\,1{,}4\,5\,8 \\
\hline
\end{array}
$$

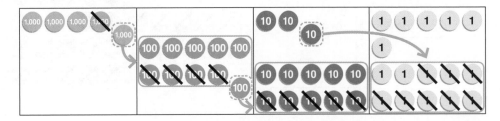

$$
\begin{array}{r}
9 \\
4\ \ 10\ 12\ 16 \\
\cancel{5}{,}\cancel{0}\ \cancel{3}\ \cancel{6} \\
-\,1{,}4\ \ 5\ \ 8 \\
\hline
\end{array}
$$

5,036 = 4,000 + 900 + 120 + 16

2 Subtract 1,568 from 5,006.

$$
\begin{array}{r}
5{,}0\,0\,6 \\
-\,1{,}5\,6\,8 \\
\hline
\end{array}
$$

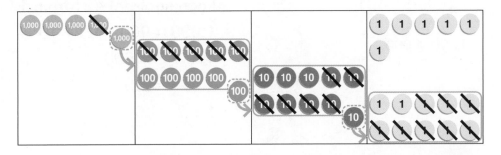

$$
\begin{array}{r}
4\ \ 9\ \ 9\ \ 16 \\
\cancel{5}{,}\cancel{0}\ \cancel{0}\ \cancel{6} \\
-\,1{,}5\ \ 6\ \ 8 \\
\hline
\end{array}
$$

5,006 = 4,000 + 900 + 90 + 16

3 Find the value of 1,082 − 985.

$$\begin{array}{r} 1{,}082 \\ -\ \ 985 \\ \hline \end{array}$$

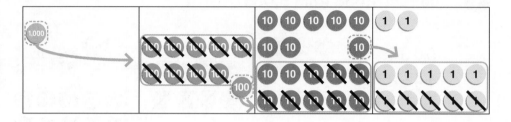

4 Subtract 421 from 3,000.

$$\begin{array}{r} 3{,}000 \\ -\ \ 421 \\ \hline \end{array}$$

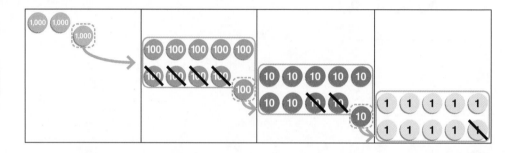

3,000 − 421
  ╱   ╲
2,000 1,000

I can subtract 421 from 1,000 mentally.
1,000 = 900 + 90 + 10
400 + ? = 900
20 + ? = 90
1 + ? = 10

5 Subtract 4,276 from 9,000.

$$\begin{array}{r} 9{,}000 \\ -\ 4{,}276 \\ \hline \end{array} \qquad \begin{array}{r} \\ +\ 4{,}276 \\ \hline \end{array}$$

Add to check the answer.

6 Find the value.

(a) 1,000 – 6 | 2,000 – 6

(b) 1,000 – 70 | 3,000 – 70

(c) 1,000 – 300 | 4,000 – 300

(d) 1,000 – 376 | 9,000 – 376

7 Find the value.

(a) 1,503 – 76

(b) 7,083 – 697

(c) 5,003 – 697

(d) 6,007 – 5,588

(e) 5,000 – 3,333

(f) 10,000 – 4,714

8 What are the missing digits?

(a)
```
   3 , 8  0  2
 -   ▨  2  ▨
   ▨ , 2  7  5
```

(b)
```
   ▨ , 0  ▨  9
 - 3 , 3  3  3
   2 , ▨  9  6
```

Exercise 3 • page 79

Think

About how much are the two items together?

About $370   About $420

How did Sofia and Dion find their estimates?

Learn

349 + 67

300 + 70 = 370

I rounded each number to the highest place.

The total cost will be about $ ___.

---

349 + 67

350 + 70 = 420

I rounded each number to the nearest ten.

The total cost will be about $ ___.

Which **estimate** is more accurate?
Which method is quicker?

Do

1 Estimate by rounding each number to the highest place.

(a) 478 + 249

$\downarrow$ $\downarrow$

500 + 200 = ⬚

(b) 184 − 58

$\downarrow$ $\downarrow$

200 − ⬚ = ⬚

(c) 725 − 294

(d) 173 + 82

2 Estimate by rounding each number to the nearest ten.

(a) 729 − 233

$\downarrow$ $\downarrow$

730 − 230 = ⬚

(b) 457 + 85

$\downarrow$ $\downarrow$

⬚ + 90 = ⬚

(c) 353 + 215

(d) 715 − 29

3 Estimate the value.

(a) 519 + 79

(b) 788 + 44

(c) 718 − 86

(d) 657 − 39

(e) 91 + 208

(f) 307 − 192

(g) 777 − 291

(h) 47 + 187 + 64

(i) 480 + 375 + 7

4 Dion wants to buy a banjo for $128 and a set of 4 recorders for $74. About how much do they cost in all?

Exercise 4 • page 83

Think

About how much more does the Solar banjo cost than the Goldenrod banjo?

Solar $4,275

Goldenrod $696

Think of different ways to round the numbers to get an estimate.

Learn

4,275 − 696
↓   ↓
4,000 − 700 = 3,300

The Solar banjo costs about $ [ ] more than the Goldenrod banjo.

4,275 − 696
↓   ↓
4,300 − 700 = 3,600

The Solar banjo costs about $ [ ] more than the Goldenrod banjo.

When might you want to make a more accurate estimate?
When might you want to make a quick estimate?

Do

1 Estimate by rounding each number to the highest place.

(a) 6,542 + 3,109

↓ ↓

7,000 + 3,000 = ____

(b) 4,462 − 537

↓ ↓

4,000 − ____ = ____

(c) 6,072 − 2,700

(d) 2,731 + 660

2 Estimate by rounding each number to the nearest hundred.

(a) 6,502 − 2,360

↓ ↓

6,500 − 2,400 = ____

(b) 4,285 + 468

↓ ↓

____ + 500 = ____

(c) 3,278 + 3,657

(d) 6,832 − 730

3 Estimate the value.

(a) 5,396 + 628

(b) 2,488 + 977

(c) 6,721 − 644

(d) 3,560 + 2,300

(e) 2,634 − 1,299

(f) 4,753 + 1,327

(g) 8,059 − 4,922

(h) 7,876 + 40

(i) 1,384 + 975 + 547

4 A school raised $2,859 at a fund drive and $735 at a bake sale. About how much money did the school raise?

Exercise 5 • page 85

Think

A store has 1,400 stands in stock.
580 of them are instrument stands.
The rest are music stands.
What is the difference in the number
of each kind of stand the store has?

Learn

Which bar should I make longer?
$1,400 - 600 = 800$
It has about 800 music stands.

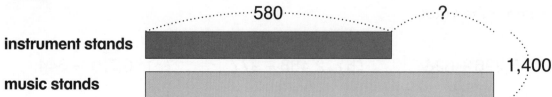

580   ?

instrument stands

music stands

1,400

I need to find the number of music stands first.

$1,400 - 580 = 820$

$820 - 580 = $ ▭

It has ▭ more music stands than instrument stands.

Do

1 Adriana sold 2,390 concert tickets.
She sold 750 fewer tickets than Jackson.
How many tickets did they sell altogether?

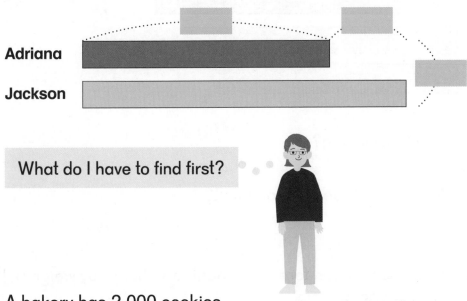

What do I have to find first?

2 A bakery has 2,000 cookies.
It sold 999 cookies in the morning and 850 in the afternoon.
How many cookies are left?

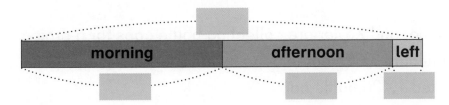

3 Imami has 840 polished rocks and 730 unpolished rocks.
She sold some rocks and now has 920 rocks left.
How many did she sell?

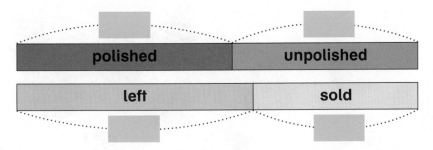

4 Aiyana jogged 975 m.
Brianna jogged 195 m farther than Aiyana.
Clara jogged 400 m less than Briana.

Aiyana	
Briana	
Clara	

(a) How far did Clara jog?

(b) How far did the three girls jog in all?

5 Together, Swazi, an African forest elephant, and her baby weigh 6,104 lb.
The baby weighs 205 lb.
How much less does the baby weigh than her mother?

6 James, Santiago, and Xavier are collecting bottle caps for a project.
They have collected 3,625 bottle caps so far.
James collected 1,430 bottle caps.
Santiago collected 182 bottle caps fewer than James.

(a) How many bottle caps did James and Santiago collect?

(b) How many fewer bottle caps did Xavier collect than James?

Exercise 6 • page 87

3-6 Word Problems

1 Estimate, then find the value.

(a) 1,575 + 783

(b) 1,309 + 3,494

(c) 8,976 + 87

(d) 6,428 + 2,539

(e) 6,742 − 964

(f) 7,204 − 1,207

(g) 8,088 − 3,509

(h) 5,001 − 94

(i) 4,008 − 1,239

(j) 9,000 − 5,855

(k) 225 + 750 + 625

(l) 1,487 + 265 + 2,277

2 Janet spent $2,080 one year on guitar lessons and $3,120 on voice lessons.

(a) How much more did the voice lessons cost than the guitar lessons?

(b) How much did she spend on both types of lessons that year?

3 The United States of America was founded on July 4, 1776.
On July 1, 1867, Canada was founded.
How many years passed between the foundation of the two countries?

4 The summit of Mount Everest is 8,850 meters above sea level.
The summit of Mount Denali is 2,670 meters lower than the summit of Mount Everest.
How high above sea level is the summit of Mount Denali in meters?

5 Liam collected 2,324 coins.
He collected 489 coins more than Camilla.
How many coins did they collect altogether?

6 Yoko saved $950.
She saved $300 more than Malik.
Jasmine saved $200 more than Malik.
How much money did Jasmine save?

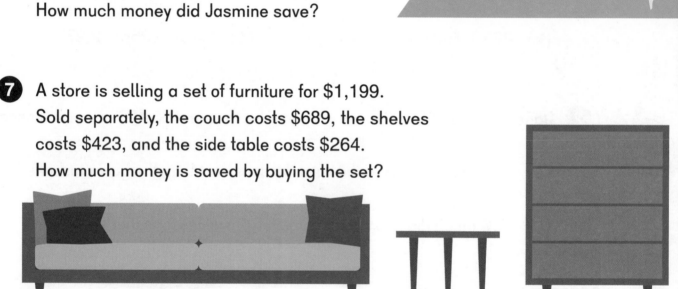

7 A store is selling a set of furniture for $1,199.
Sold separately, the couch costs $689, the shelves costs $423, and the side table costs $264.
How much money is saved by buying the set?

8 Heather wants to buy a game console that costs $314.
If she buys it online, it will cost $285.
Estimate the amount of money she will save if she buys it online.

Exercise 7 • page 91

3-7 Practice

Chapter 4

Multiplication and Division

I want to make 6 flowers.
Each flower has 3 pipe cleaners.
How many pipe cleaners do I need?
6 groups of 3 is 6 × 3.

Make up other multiplication expressions.

Think

What multiplication expressions could
we write to find the total number of pompoms?

Learn

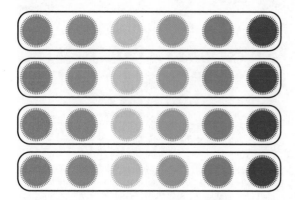

4 groups of 6
4 **times** 6
4 × 6

6 + 6 + 6 + 6 =

4 × 6 =

We can also think of this as
6 in the group **multiplied by** 4.
6 × 4

We multiply to find the total when we are adding equal groups.
× is the symbol used for **multiplication**.
It means **times** or **multiplied by**.

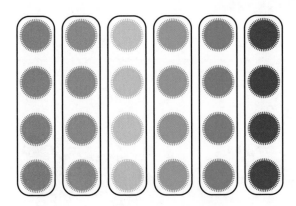

I see 6 **times** 4.
6 × 4

I see a group of 4 **multiplied by** 6.
4 × 6

4 + 4 + 4 + 4 + 4 + 4 =

6 × 4 =

4 groups of 6 has the same number of objects as 6 groups of 4.

The **product** of 6 and 4 is 24.
The product of 4 and 6 is 24.

There are   pompoms.

Do

1 (a) How many ants are in 3 groups of 6?

$3 \times 6 = $

(b) How many ants are in 6 groups of 3?

$6 \times 3 = $

2 (a) Multiply 9 by 4.

$9 \times 4 = $

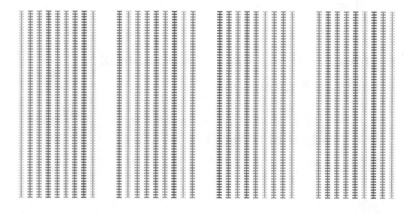

(b) Multiply 4 by 9.

$4 \times 9 = $

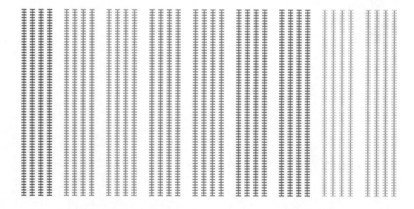

3 How many garlic heads are there?

3 × 10 = ⬚ │ 10 × 3 = ⬚

4 What is the product of 2 and 4?

2 × 4 = ⬚ │ 4 × 2 = ⬚

5 There are 8 children.
Each child has 3 balloons.
Find the total number of balloons.

⬚ × ⬚ = ⬚

6 There are 3 cards.
Each card has 8 stickers on it.
Find the total number of stickers.

⬚ × ⬚ = ⬚

Exercise 1 • page 95

Think

Find the missing products.
What patterns do you notice?

How can we use the facts we know to find the facts we don't know?

×	1	2	3	4	5	10
1	1	2	3	4	5	10
2	2	4	6		10	20
3	3	6	9	12	15	
4	4		12		20	40
5	5	10	15	20	25	50
6	6	12			30	60
7	7		21			70
8	8	16			40	80
9	9	18				90
10	10	20		40	50	100

I know 5 × 4 and 2 × 4.
Can I use that to find 7 × 4?

5 × 3 = 15
What is 6 × 3?

Since 8 × 2 = 16, 8 × 4 = ?

Learn

$5 \times 3 = 15$

$1 \times 3 = 3$

$6 \times 3 = $

$6 \times 3 = 15 + ?$

5   1

$5 \times 4 = 20$

$2 \times 4 = $

$7 \times 4 = $

$7 \times 4 = 20 + ?$

5   2

$8 \times 2 = 16$

$8 \times 2 = 16$

$8 \times 4 = $

$8 \times 4 = 16 + 16$

2   2

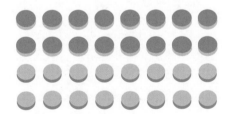

If I know 8×4, I also know 4×8.

Do

1 10 × 4 = ☐

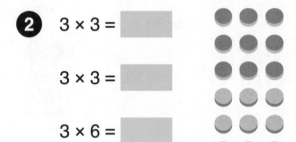

1 × 4 = ☐

9 × 4 = ☐

$9 \times 4 = 40 - ?$

4 × 9 = ☐

2 3 × 3 = ☐

3 × 3 = ☐

3 × 6 = ☐

3 7 × 5 is 2 × 5 more than ☐ × 5.

7 × 5 = ☐

4 9 × 3 is ☐ × 3 less than 10 × 3.

9 × 3 = ☐

5 (a) 4 × 4 = ☐

(b) 5 × 9 = ☐

(c) 2 × 7 = ☐

(d) 10 × 4 = ☐

(e) 4 × 6 = ☐

(f) 3 × 8 = ☐

Exercise 2 • page 98

4-2 Strategies for Finding the Product

Think

There are 15 brushes.

(a) Put the brushes equally into 3 boxes.
    How many brushes are in each box?

(b) Put 3 brushes into each box.
    How many boxes are needed?

Learn

(a)

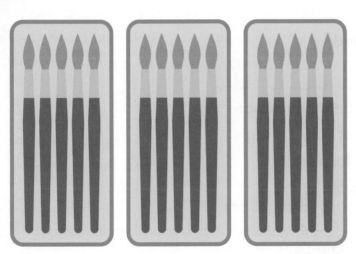

$$15 \div 3 = 5$$

↑   ↑   ↑

total   number   number in
      of groups   each group

Each box gets ▢ brushes.

We are sharing the brushes equally into 3 boxes.
$3 \times ? = 15$

(b)

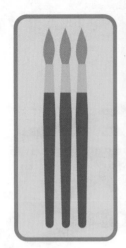 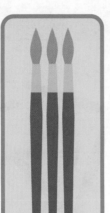  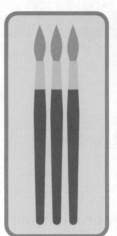 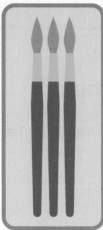

$$15 \div 3 = 5$$

↑ total   ↑ number in each group   ↑ number of groups

 boxes are needed.

We are grouping the brushes by 3.
? × 3 = 15

We divide to find the amount in each group or the number of equal groups.
÷ is the symbol used for **division**.
We read ÷ as **divided by**.

The two situations are different, but the expression for both is 15 ÷ 3 and the answer for both is 5.

The **quotient** of 15 divided by 3 is 5.

Do

1 Divide 21 counters into 3 groups.
How many counters are in each group?

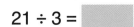

Put 21 counters into groups of 3.
How many groups are there?

$21 \div 3 = $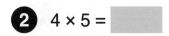

To divide 21 by 3, we can use the multiplication facts of 3,
$3 \times ? = 21$ or $? \times 3 = 21$.

2 $4 \times 5 = $ | $20 \div 4 = $

$5 \times 4 = $ | $20 \div 5 = $

3

$2 \times 5 = $ | $10 \div 2 = $

$5 \times 2 = $ | $10 \div 5 = $

4 (a) $4 \times $ $= 28$ | $28 \div 4 = $

(b) $5 \times $ $= 45$ | $45 \div 5 = $

(c) $\times 3 = 24$ | $24 \div 3 = $

(d) $\times 10 = 90$ | $90 \div 10 = $

5 (a) $28 \div 4 =$ [ ] (b) $27 \div 3 =$ [ ] (c) $16 \div 2 =$ [ ]

(d) $35 \div 5 =$ [ ] (e) $16 \div 4 =$ [ ] (f) $80 \div 10 =$ [ ]

(g) $36 \div 4 =$ [ ] (h) $25 \div 5 =$ [ ] (i) $32 \div 4 =$ [ ]

6 Ximena has 45 pennies.
She wants to trade them in for nickels.
How many nickels will she get?

$? \times 5 = 45$

$45 \div 5 =$ [ ]

She will get [ ] nickels.

7 Caleb has some yarn that is 30 m long.

(a) If he cuts it into 3 equal pieces, how long will each piece be?

(b) If he cuts it into pieces that are each 3 m long,
how many pieces will he have?

8 There are 12 red pens, 20 blue pens, and 4 bags.
Paul put an equal number of each color of pen in each bag.

(a) How many pens of each color are in each bag?

(b) How many pens are there in each bag?

Exercise 3 • page 101

4-3 Looking Back at Division

Think

3 children are sharing some glue sticks.

(a) How many glue sticks will each child get if there are 6 glue sticks?

(b) How many glue sticks will each child get if there are 3 glue sticks?

(c) How many glue sticks will each child get if there are 0 glue sticks?

Learn

(a)

$3 \times \boxed{} = 6$ | $6 \div 3 = \boxed{}$

Each child gets $\boxed{}$ glue sticks.

(b)

$3 \times \boxed{} = 3$ | $3 \div 3 = \boxed{}$

Each child gets $\boxed{}$ glue stick.

(c)

$3 \times \boxed{} = 0$ | $0 \div 3 = \boxed{}$

Each child gets $\boxed{}$ glue sticks.

If we multiply 3 by 0, the product is 0.

If we divide 0 by 3, there are 0 in each group.

Do

1 How many birds are there?

(a)  $3 \times 2 =$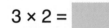

(b) $3 \times 1 =$

(c) $3 \times 0 =$

(d) $2 \times 0 =$

(e) $1 \times 0 =$

(f) 0 branches with 0 birds $0 \times 0 =$

The product of any number and 0 is 0.

2 (a) Put 6 counters into 2 equal groups.

$2 \times \boxed{} = 6 \quad | \quad 6 \div 2 = \boxed{}$

(b) Put 6 counters into 1 group.

$1 \times \boxed{} = 6 \quad | \quad 6 \div 1 = \boxed{}$

What happens when we divide a number by 1?

(c) Why can't we put 6 counters into 0 groups?

We cannot divide a number by 0.

3 (a) Divide 8 counters into 8 equal groups.

$8 \div 8 = \boxed{}$

(b) Divide 5 counters into 5 equal groups.

$5 \div 5 = \boxed{}$

What happens when we divide any number by itself?

4 (a) Divide 0 counters into 10 equal groups.

$$0 \div 10 = \boxed{}$$

(b) Divide 0 counters into 5 equal groups.

$$0 \div 5 = \boxed{}$$ 0 divided by any number except 0 is...

5 Karen has 8 crayons.
How many boxes does she need to put...

(a) 4 crayons in each box?   (b) 2 crayons in each box?

(c) 1 crayon in each box?   (d) 8 crayons in each box?

6 (a) $0 \times 9 = \boxed{}$ (b) $7 \times 0 = \boxed{}$ (c) $0 \times 0 = \boxed{}$

(d) $0 = 5 \times \boxed{}$ (e) $\boxed{} \times 12 = 0$ (f) $0 = 20 \times \boxed{}$

7 (a) $1 \div 1 = \boxed{}$ (b) $0 \div 9 = \boxed{}$ (c) $7 \div 1 = \boxed{}$

(d) $5 \div \boxed{} = 5$ (e) $8 \div \boxed{} = 1$ (f) $\boxed{} \div 10 = 0$

8 (a) $10 \div 10 = 8 \div \boxed{}$ (b) $0 \div 4 = \boxed{} \div 7$

(c) $0 \div 1 = \boxed{} \times 3$ (d) $4 \times 0 = \boxed{} \times 6$

Exercise 4 • page 104

Think

Mei has 14 pipe cleaners.
She wants to make flowers.
Each flower needs 3 pipe cleaners.
How many flowers can she make?

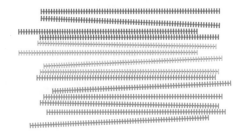

Learn

$14 \div 3$ is 4 with a remainder of 2.
$14 = 4 \times 3 + 2$

$4 \times 3 = 12$,
which is almost 14.
$14 - 12 = 2$

We can use R to stand for **with a remainder of**.
$14 \div 3$ is 4 R 2.
4 is the quotient.
The remainder 2 was not divided.

She can make [　] flowers.

There are [　] pipe cleaners left over.

quotient   remainder
   ↓   ↓
[　] × 3 + [　] = 14   Check the answer to $14 \div 3$.

Do

1 (a) Put 6 counters into groups of 3. $6 \div 3 =$ ___

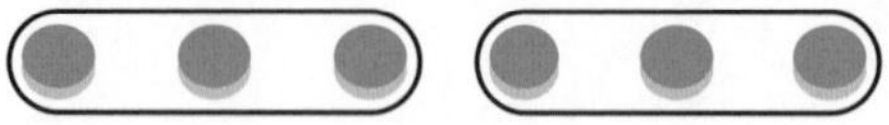

(b) Put 7 counters into groups of 3. $7 \div 3$ is ___ with a remainder of ___.

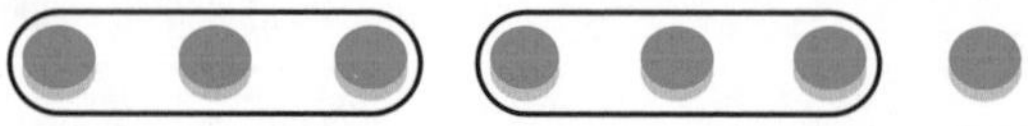

(c) Put 8 counters into groups of 3. $8 \div 3$ is ___ with a remainder of ___.

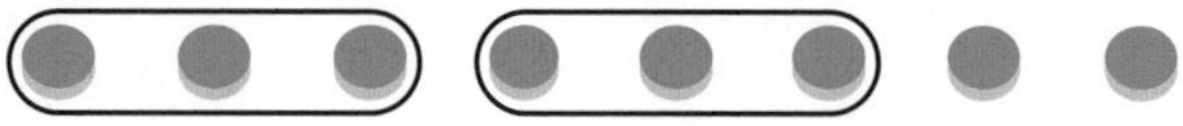

(d) Put 9 counters into groups of 3. $9 \div 3 =$ ___

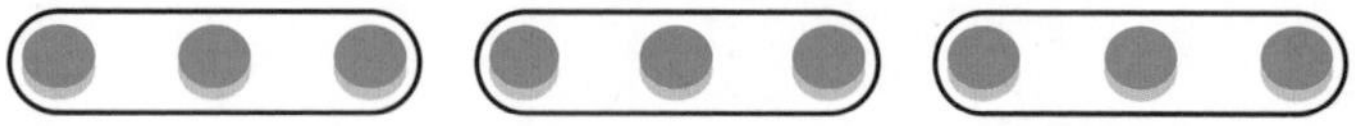

When we divide by 3,
the remainder must be less than 3.

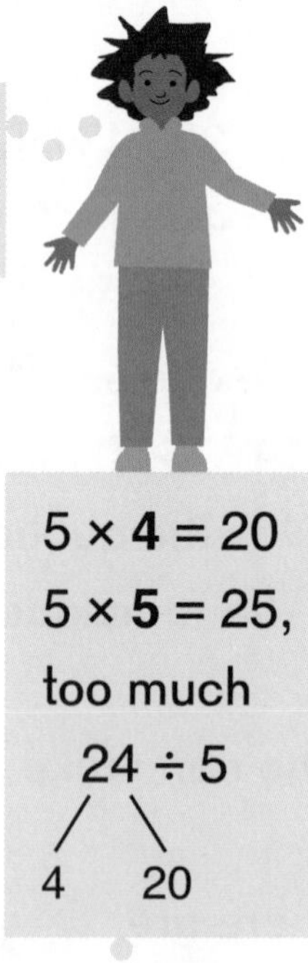

2 5 children are sharing 24 googly eyes equally.
How many eyes does each child get?
How many are left over?

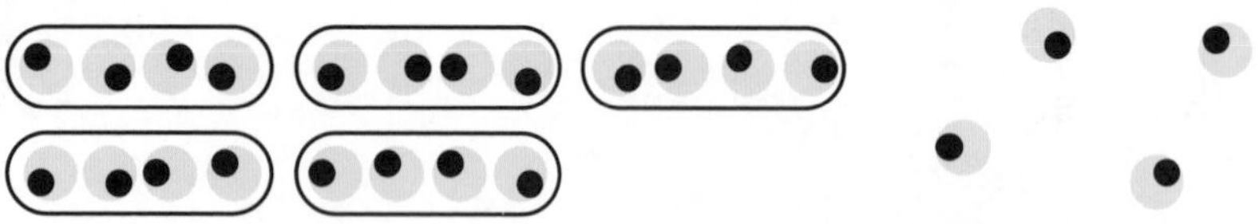

$5 \times \mathbf{4} = 20$
$5 \times \mathbf{5} = 25$, too much

$24 \div 5$ → 4, 20

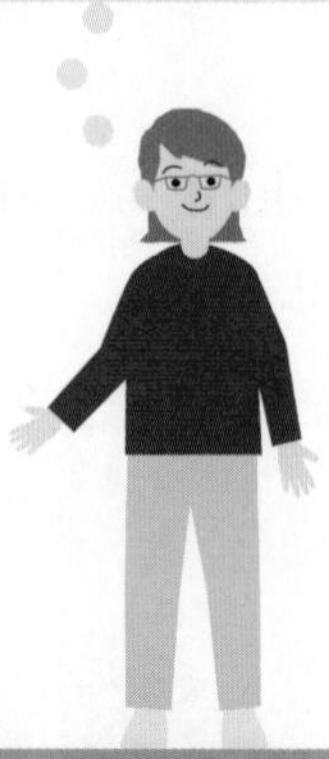

$24 \div 5$ is ___ with a remainder of ___.

Each child gets ___ googly eyes.

___ googly eyes are left over.

3 (a) 25 ÷ 5 =

(b) 26 ÷ 5 is [   ] with a remainder of [   ].

(c) 27 ÷ 5 is [   ] with a remainder of [   ].

(d) 28 ÷ 5 is [   ] with a remainder of [   ].

(e) 29 ÷ 5 is [   ] with a remainder of [   ].

(f) 30 ÷ 5 = [   ]

(g) 31 ÷ 5 is [   ] with a remainder of [   ].

When we divide by 5, the remainder has to be less than...

4 Find the quotient and remainder.
Check your answers.

(a) 15 ÷ 2

(b) 28 ÷ 5

(c) 29 ÷ 3

(d) 30 ÷ 4

(e) 82 ÷ 10

(f) 21 ÷ 2

5 46 people are going on a trip.
Each van can hold 10 people.
What is the fewest number of vans they will need?

6 Pamela has 38 pencils and 4 pencil holders.
What is the greatest number of pencils that can be put in
each holder so that each has the same number of pencils?

Exercise 5 • page 107

Think

There are ☐ erasers.

We want to put 2 erasers in each pencil case.

Find the number of pencil cases for ☐6☐, ☐7☐, ☐8☐, ☐9☐, and ☐10☐ erasers.

Which of the numbers can be divided **evenly** by 2?

Which of the numbers cannot be divided **evenly** by 2?

Learn

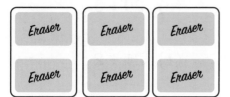

There is no remainder so
6 can be divided **evenly** by 2.

6 ÷ 2 = ☐

There is a remainder,
so 7 cannot be
divided evenly by 2.

7 ÷ 2 is ☐ with a remainder of ☐.

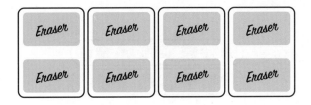

$8 ÷ 2 = $ ⬜

8 and 10, but not 9, can be divided evenly by 2.

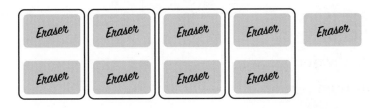

$9 ÷ 2$ is ⬜ with a remainder of ⬜.

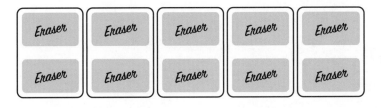

$10 ÷ 2 = $ ⬜

What other numbers can be divided evenly by 2?

Even numbers can be divided by 2 with no remainder.
Odd numbers have a remainder of 1 when divided by 2.

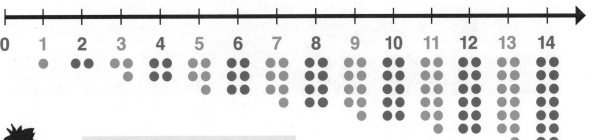

$0 ÷ 2 = 0$, so 0 is an even number.

Do

1 (a) Put 16 counters into 2 groups.
Is 16 even or odd?

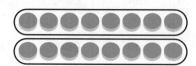

(b) Put 19 counters into 2 groups.
Is 19 even or odd?

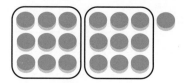

An even number of
objects can be divided
by 2 with no remainder.

2 Divide each number by 2.
Is the number even or odd?

(a) 17   (b) 20   (c) 15   (d) 18

3 19 people are camping in tents.
2 people can sleep in each tent.

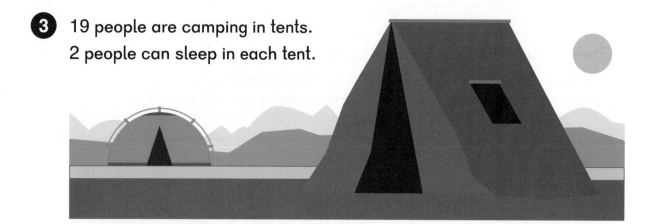

(a) Are there an odd or even number of campers?

(b) What is the least number of tents needed?

4 Are the sums even or odd?

(a) 4 + 6 =

(b) 3 + 5 =

(c) 6 + 3 =

(d) 5 + 6 =

Can I tell by looking at the two numbers whether their sum or product will be odd or even?

5 Are the products even or odd?

(a) 4 × 4 =

(b) 3 × 5 =

(c) 4 × 3 =

(d) 3 × 6 =

Exercise 6 • page 111

Think

Sofia, Alex, and Emma are making cactus pots.
They are painting flat rocks to look like cacti and putting them in pots.

(a) Sofia has 3 pots.
She painted 6 rocks for each pot.
How many rocks did she use?

How are the problems similar?
How are they different?

(b) Alex has 3 pots.
He used 18 rocks altogether.
He put the same number of rocks in each pot.
How many rocks are in each pot?

(c) Emma used 18 rocks.
She put 3 rocks in each of her pots.
How many pots did she use?

Draw a bar model and write an expression for each problem.
Solve each problem.

<u>Learn</u>

(a)

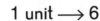
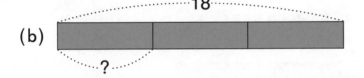

1 unit ⟶ 6
3 units ⟶ 3 × 6 = ▢
Sofia used ▢ rocks.

We have to
find the total.

(b)

18

?

3 units ⟶ 18
1 unit ⟶ 18 ÷ 3 = ▢
Alex put ▢ rocks in each pot.

We know the total.
We have to find
the value of 1 unit.

(c)

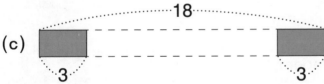

18

3   3

1 unit ⟶ 3
▢ units ⟶ ▢ × 3 = 18
18 ÷ 3 = ▢

Emma used ▢ pots.

We know the total.
We have to find the
number of units.

All three problems have
equal groups.
A **unit** is an equal group.

Do

1 What are the missing numbers?

(a)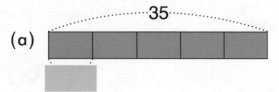

(b)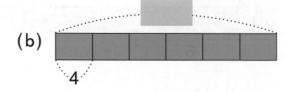

2 Dion made 4 apple trees.

He used 36 pompoms for the apples.

He put the same number of pompoms on each tree.

How many pompoms did he put on each tree?

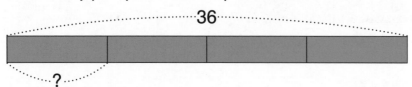

4 units ⟶ 36

1 unit ⟶ 36 ◯ ▢ = ▢

What sign goes in the ◯ ?

Dion put ▢ pompoms on each tree.

3 Each panda needs 5 black pompoms.

Sofia has 45 black pompoms.

How many pandas can she make?

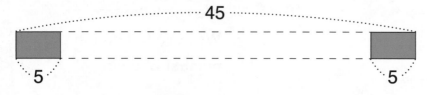

45 ◯ ▢ = ▢

Sofia can make ▢ pandas.

4 Mei made 4 spiders.
She used 8 craft sticks for each spider.
How many craft sticks did she use?

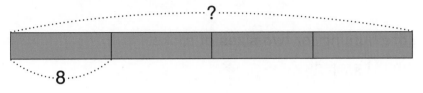

1 unit ⟶ 8
4 units ⟶ ▮ ◯ 8 = ▮

Mei used ▮ craft sticks.

5 Draw models and solve.

(a) There are 4 boxes of decorations.
Each box has 10 decorations in it.
How many decorations are there?

(b) Renata has 20 ft of yarn.
She cuts the yarn into pieces each 2 ft long.
How many pieces does she have?

(c) Isaac used 27 pieces of ribbon to make 3 windsocks.
Each windsock has the same number of ribbons.
How many pieces of ribbon did he use for each windsock?

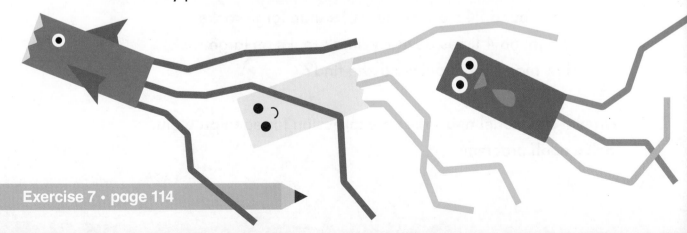

Exercise 7 • page 114

Think

Mei, Dion, and Sofia are hunting for two sizes of rocks,
large and small, for an art project.

(a) Mei found 32 small rocks.
     She found 4 times as many small rocks as large rocks.
     How many large rocks did she find?

(b) Dion found 8 large rocks.
     He found 4 times as many small rocks as large rocks.
     How many rocks did he find in all?

(c) Sofia found 24 more small rocks than large rocks.
     She found 4 times as many small rocks as large rocks.
     How many large rocks did she find?

Draw a bar model and write an expression for each problem.
Solve each problem.

Learn

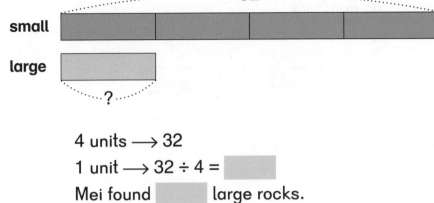

(a) **small**

large

4 units ⟶ 32
1 unit ⟶ 32 ÷ 4 = ▢
Mei found ▢ large rocks.

(b) **small**

large

1 unit ⟶ 8
5 units ⟶ 5 × 8 = ▢
Dion found ▢ rocks in all.

There are 5 units **in all**.

(c) **small**

large

3 units ⟶ 24
1 unit ⟶ 24 ÷ 3 = ▢
Sofia found ▢ large rocks.

There are 3 **more** units of small than large rocks.

Do

1 What are the missing numbers?

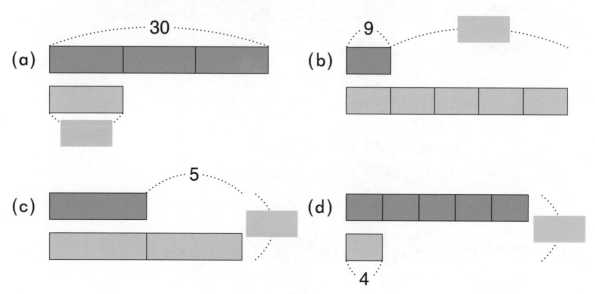

(a)

(b)

(c)

(d)

2 Alex made 3 times as many pipe-cleaner snakes as monkeys.

He made 12 snakes.

How many monkeys did he make?

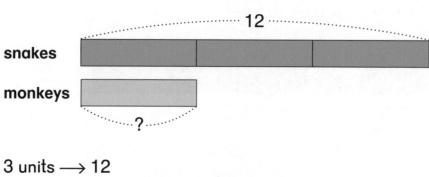

snakes

monkeys

3 units ⟶ 12

1 unit ⟶ 12 ◯ ⬜ = ⬜

He made ⬜ monkeys.

3 A set of brushes costs $5.

An acrylic paint set costs 5 times as much as the brushes.

How much more does the paint set cost than the brushes?

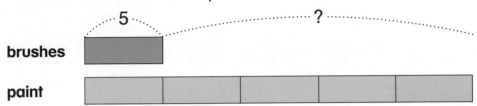

1 unit ⟶ 5

4 units ⟶ 4 ◯ ▨ = ▨

The paint set costs $ ▨ more than the brushes.

4 Draw models and solve.

(a) Susma has 9 sticks of green modeling clay.
   She has 3 times as many sticks of green clay as red clay.
   How many sticks of red clay does she have?

(b) Franco has 9 polished rocks.
   He has 3 times as many unpolished rocks as polished rocks.
   How many unpolished rocks does he have?

(c) Valentina and Mayam collected 20 bottle caps for an art project.
   Mayam collected 3 times as many as Valentina.
   How many bottle caps did Valentina collect?

(d) James has $20 less than Bron.
   Bron has 3 times as much money as James.
   How much money does James have?

Exercise 8 • page 116

Think

Mei had 30 m of ribbon.

She cut off 2 pieces of ribbon.

The second piece is 3 times as long as the first piece.

There is still 18 m of ribbon left on the spool.

How long is each piece?

I need to find the total length of the two cut pieces first.

Learn

first piece	
second piece	
left on spool	

30

18

4 units ⟶ 30 − 18 = 12

1 unit ⟶ 12 ÷ 4 = 3

The first piece is ⬚ m long.

3 units ⟶ 3 × 3 = ⬚

The second piece is ⬚ m long.

Check your answers.
Does 3 + 9 + 18 = 30?

Do

1 Mei made 3 times as many ants as spiders.
She made 12 ants.

(a) How many animals did she make?

(b) How many more ants than spiders did she make?

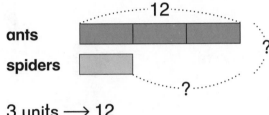

Find the value
of 1 unit first
and use that for
both problems.

3 units ⟶ 12
1 unit ⟶ 12 ÷ 3 = 4

(a) 4 units ⟶ 4 × ▢ = ▢ │ She made ▢ animals.

(b) 2 units ⟶ 2 × ▢ = ▢ │ She made ▢ more ants than spiders.

2 Ms. Davis bought 6 skeins of wool yarn for
$5 each and a set of knitting needles for $12.
How much did she spend?

Find the cost of
the yarn first.

1 unit ⟶ 5
6 units ⟶ 6 ◯ ▢ = ▢ (cost of yarn)

▢ + 12 = ▢ (total spent)

She spent $ ▢ .

3 Alex and Emma together made 23 dinosaurs.
Alex made 5 more dinosaurs than Emma.
How many dinosaurs did Emma make?

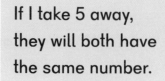

If I take 5 away, they will both have the same number.

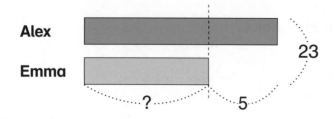

Alex

Emma

? 5 23

We need to find how many Emma made.
Make her bar the unit.

2 units ⟶ 23 − 5 = 18
1 unit ⟶ 18 ÷ 2 =

Emma made ___ dinosaurs.

4 Dexter bought 3 packs of foam brushes.
There were 4 thin brushes and 2 thick
brushes in each pack.
How many brushes did he buy?

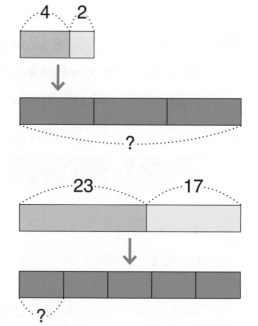

4   2

?

5 Sita polished 23 rocks on Monday
and 17 rocks on Tuesday.
She put the rocks equally into 5 boxes.
How many rocks are in each box?

23   17

?

6 Mei and Dion together made 11 turtles.
Mei made 3 more turtles than Dion.
How many turtles did Mei make?

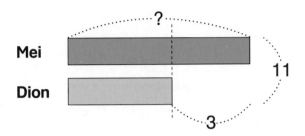

Make Mei's bar the unit.
If Dion had made 3 more
then...

7 Asimah has 9 tulips.
She has 3 times as many daisies as tulips.
She arranges 6 flowers in each vase.
How many vases does she use?

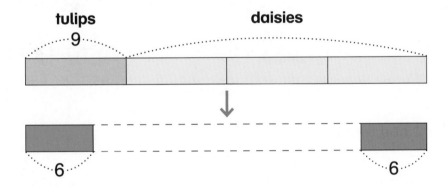

8 A pack of 5 paint pens cost $3.
Mr. Ikeda bought 20 paint pens.
How much did he pay?

9 Hudson has 4 times as many crayons as Elena.
He has 24 more crayons than Elena does.
How many crayons do they have altogether?

Exercise 9 • page 119

1 Find the value.

(a) $8 \div 4$

(b) 4×7

(c) $14 \div 2$

(d) 4×4

(e) $32 \div 4$

(f) 0×10

(g) $35 \div 5$

(h) $27 \div 3$

(i) $18 \div 3$

(j) $16 \div 2$

(k) $5 \div 5$

(l) $0 \div 10$

2 (a) $5 \times \underline{} = 20$

(b) $\underline{} \times 5 = 0$

(c) $\underline{} = 8 \times 3$

(d) $3 \div \underline{} = 3$

(e) $\underline{} \div 5 = 0$

(f) $\underline{} = 4 \div 2$

3 Find the quotient and remainder.

(a) $7 \div 2$

(b) $10 \div 3$

(c) $22 \div 4$

(d) $16 \div 5$

(e) $42 \div 10$

(f) $88 \div 10$

(g) $26 \div 3$

(h) $26 \div 4$

(i) $26 \div 5$

4 Are the following numbers odd or even?

(a) 12

(b) 11

(c) 13

(d) 16

5 Katie is making decorative balls out of yarn to sell at the farmer's market on Kids Vending Day.

She bought 1 pack of balloons, 10 skeins of yarn, and 2 bottles of glue.

She spent $10 on the pack of balloons.

The balloons cost 5 times as much as 1 skein of yarn.

The 2 bottles of glue cost the same as 3 skeins of yarn.

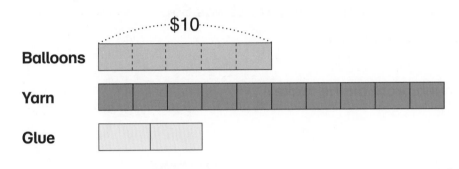

(a)   How much does 1 skein of yarn cost?

(b)   How much did she spend on the yarn?

(c)   How much does 1 bottle of glue cost?

(d)   How much did she spend in all?

(e)   Katie made 9 each of red, yellow, orange, and green balls.
      She made 4 brown balls.
      She displayed the balls by putting 5 in each bowl.
      How many bowls did she use?

(f)   She sold all 9 red balls.
      One buyer gave her $1 extra as a tip.
      She received $28 from selling the red balls.
      How much did she sell each red ball for?

6 Josef made a total of 5 birdhouses and sold them all for $9 each at the market on Kids Vending Day.

(a) How much money did he receive?

(b) The materials for each birdhouse cost $3.
The fee for the booth at the market was $5.
How much did he spend?

(c) How much profit did he make?

7 Evan collected 2 more pinecones than Arman. Mila collected twice as many pinecones as Evan. Altogether, they collected 30 pinecones.

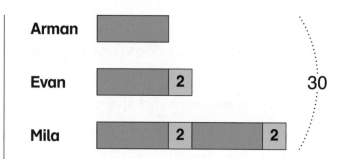

(a) How many pinecones did Arman collect?

(b) How many pinecones did Mila collect?

8 Alisha collected 3 more pinecones than Fuyu.
Lucas collected 3 times as many pinecones as Alisha.
Altogether, they collected 47 pinecones.

(a) How many pinecones did Fuyu collect?

(b) How many pinecones did Lucas collect?

Exercise 10 • page 123

① In the number 2,846...

(a) Which digit is in the thousands place?
(b) What is the value of the digit 8?
(c) In what place is the digit 4?
(d) What number is 1,000 less than this number?
(e) What number is 900 more than this number?
(f) What number is 6 tens less than this number?

②

| 4,189 | 3,899 | 4,086 | 3,914 |

(a) In which of these numbers is the digit in the tens place 1 more than the digit in the hundreds place?
(b) Arrange the numbers in order from least to greatest.
(c) Find the sum of the least and greatest of these numbers.
(d) Find the difference between the least and greatest of these numbers.

③ Round 4,549...

(a) To the nearest ten.
(b) To the nearest hundred.
(c) To the nearest thousand.

4 What sign, >, <, or =, goes in the ◯?

(a) 5,000 + 500 + 85 ◯ 5,000 + 5 + 85

(b) 8 + 40 + 600 + 3,000 ◯ 2,000 + 6,000 + 400 + 8

(c) 210 tens ◯ 21 hundreds

(d) 3 thousands + 65 hundreds ◯ 9 thousands + 50 tens

5 (a) 38 + 47 = ▢ − 5   (b) 82 − 25 = ▢ + 7

(c) 740 − 90 = ▢ + 10   (d) 475 + 98 = ▢ − 2

(e) 60 + ▢ = 100 − 32   (f) 2,000 − 480 = 1,000 + ▢

(g) 6 × 3 = 15 + ▢   (h) 4 × 9 = 40 − ▢

(i) 6 × 6 = ▢ × 4   (j) 2 × 4 = 16 ÷ ▢

(k) 6 × 0 = ▢ ÷ 6   (l) 26 = 4 × ▢ + 2

6 Estimate, then find the value.

(a) 3,875 + 283   (b) 7,004 − 2,207

(c) 6,145 − 724   (d) 2,345 + 2,655

7 Find the quotient and remainder.

(a) $9 \div 2$

(b) $19 \div 3$

(c) $30 \div 4$

(d) $48 \div 5$

(e) $94 \div 10$

(f) $13 \div 4$

8 There are 3,326 boys and 5,807 girls in a school.

(a) How many more girls than boys are there?

(b) How many children are there altogether?

9 There are 2,500 books in a library.
972 are fiction books.
There are 113 fewer non-fiction than fiction books.
The rest are reference books.

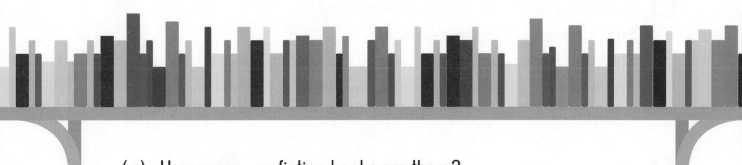

(a) How many non-fiction books are there?

(b) How many reference books are there?

10 Clemens has 47 flowers.
He wants to put the same number of flowers in each of 5 vases.
What is the least number of flowers he will have left over?

11 Xavier's dog weighs 36 lb.
His dog weighs 4 times as much as his cat.
What is the total weight of both animals?

12 Diego is making airplanes.
Each airplane uses 3 craft sticks and 1 clothespin.
Diego has 28 craft sticks and 12 clothespins.
He makes as many airplanes as he can.
How many airplanes did he make?

13 Fatima had 3 times as many bracelets as Hannah.
Then, Fatima lost 3 bracelets.
The girls now have 17 bracelets altogether.
How many bracelets does Hannah have?

14 Allison bought 24 sticks of modeling clay.
She put them equally into 4 bags.
She used the clay in 2 of the bags to make something.
How many sticks of clay does she have left?

15 How many different ways can 12 children
be divided equally into more than 1 team?
Each team must have at least 2 children.

Exercise 11 • page 127

Review 1

Chapter 5

Multiplication

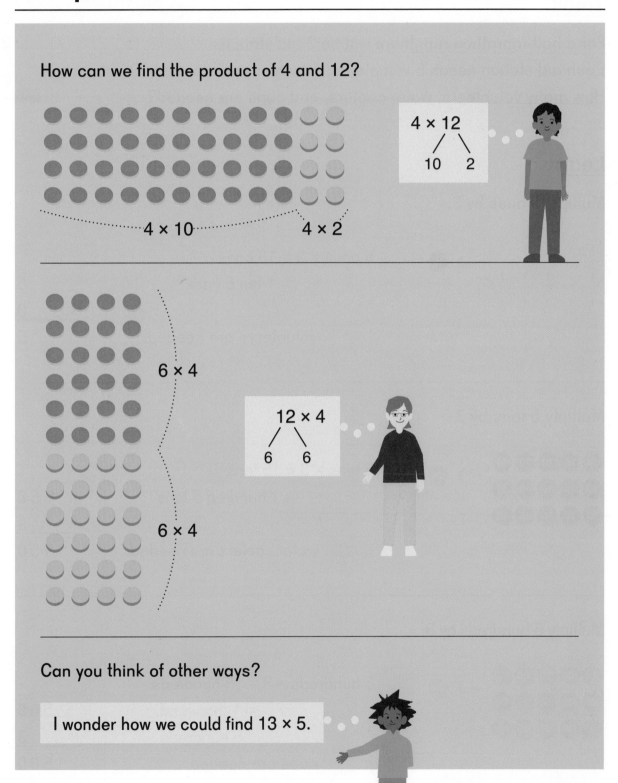

How can we find the product of 4 and 12?

4 × 12

4 × 10   4 × 2

10   2

6 × 4

12 × 4

6   6

6 × 4

Can you think of other ways?

I wonder how we could find 13 × 5.

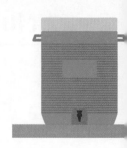

Think

For a half-marathon run, there will be 3 aid stations.
Each aid station needs 5 volunteers, 50 water coolers, and 500 cups.
How many volunteers, water coolers, and cups are needed?

Learn

Multiply 5 ones by 3.

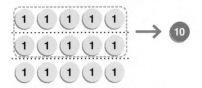

5 ones × 3 = 15 ones
= 1 ten 5 ones

[   ] volunteers are needed.

$$\begin{array}{r} 5 \\ \times\ \ 3 \\ \hline 1\,5 \end{array}$$

Multiply 5 tens by 3.

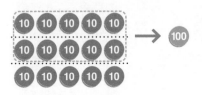

5 tens × 3 = 15 tens
= 1 hundred 5 tens

[   ] water coolers are needed.

$$\begin{array}{r} 5\,0 \\ \times\ \ \ 3 \\ \hline 1\,5\,0 \end{array}$$

Multiply 5 hundreds by 3.

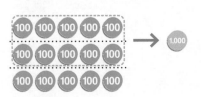

5 hundreds × 3 = 15 hundreds
= 1 thousand,
5 hundreds

[   ] cups are needed.

$$\begin{array}{r} 5\,0\,0 \\ \times\ \ \ \ 3 \\ \hline 1{,}5\,0\,0 \end{array}$$

Do

1 (a) Multiply 2 ones by 4.

2 ones × 4 =   ones

2 × 4 =

$$\begin{array}{r} 2 \\ \times\ 4 \\ \hline \end{array}$$

(b) Multiply 2 tens by 4.

2 tens × 4 =   tens

20 × 4 =

$$\begin{array}{r} 20 \\ \times\ 4 \\ \hline \end{array}$$

(c) Multiply 2 hundreds by 4.

2 hundreds × 4 =   hundreds

200 × 4 =

$$\begin{array}{r} 200 \\ \times\ 4 \\ \hline \end{array}$$

2 (a) Multiply 8 ones by 3.

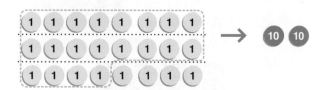

8 ones × 3 = 24 ones

= 2 tens [   ] ones

$8 \times 3 =$ [   ]

$$\begin{array}{r} 8 \\ \times\ \ 3 \\ \hline \end{array}$$

(b) Multiply 8 tens by 3.

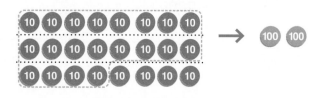

8 tens × 3 = [   ] tens

= [   ] hundreds [   ] tens

$80 \times 3 =$ [   ]

$$\begin{array}{r} 80 \\ \times\ \ 3 \\ \hline \end{array}$$

(c) Multiply 8 hundreds by 3.

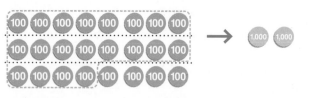

8 hundreds × 3 = [   ] hundreds

= [   ] thousands [   ] hundreds

$800 \times 3 =$ [   ]

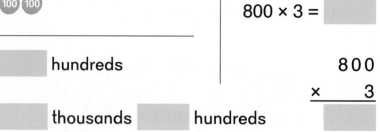

$$\begin{array}{r} 800 \\ \times\ \ 3 \\ \hline \end{array}$$

3 Multiply 4 by 500.

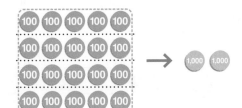

4 × 500 = 500 × 4
We can think of the answer to 4 groups of 500 to find the answer to 500 groups of 4.

$$\begin{array}{r} 5\textbf{00} \\ \times \quad 4 \\ \hline 2,0\textbf{00} \end{array}$$

20 hundreds = 2 thousand

4 × 500 = ▭

4 Find the product.

(a) 3 × 5   (b) 30 × 5   (c) 300 × 5

(d) 8 × 4   (e) 80 × 4   (f) 800 × 4

(g) 5 × 7   (h) 5 × 70   (i) 5 × 700

5

There are 5 boxes of safety pins for the race bibs.
There are 500 safety pins in each box.
How many safety pins are there?

Exercise 1 • page 133

Think

There are 3 boxes of race shirts on a table.
There are 32 shirts in each box.
How many shirts are there?

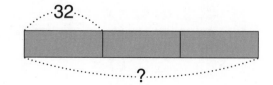

Learn

32 × 3

We can write 3 groups of
32 as 32 multiplied by 3.
3 × 32 = 32 × 3

Method 1 ►

3 tens × 3 = 9 tens	2 ones × 3 = 6 ones
30 × 3 = **90**	2 × 3 = 6

32 × 3 = 90 + 6
     = 96

32 × 3
  /   \
30   2

There are ▮▮▮▮ shirts.

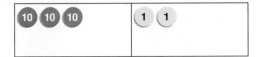

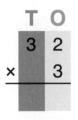

Multiply 2 ones by 3.

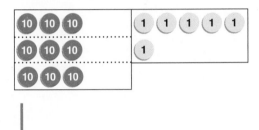

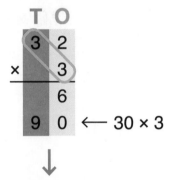

← 2 × 3

Multiply 3 tens by 3.

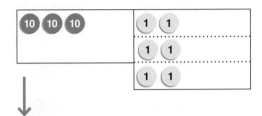

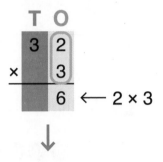

← 30 × 3

Add the products.

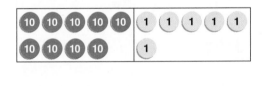

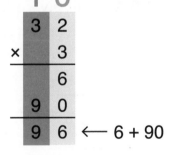

← 6 + 90

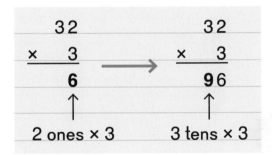

	32			32
×	3		×	3
	6	→		**9**6
	↑			↑
2 ones × 3			3 tens × 3	

I wrote the product from multiplying the tens right away.

Do

1 Find the product of 4 and 12.

$10 \times 4 = 40$

$2 \times 4 = 8$

$12 \times 4 = 40 + 8$

$ = \boxed{}$

12 × 4

10   2

2 Multiply 143 by 2.

$100 \times 2 = \boxed{}$ | $40 \times 2 = \boxed{}$ | $3 \times 2 = \boxed{}$

$143 \times 2 = \boxed{}$

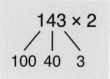

143 × 2

100 40   3

3 Multiply 23 by 3.

$$\begin{array}{r} 2\,3 \\ \times \quad 3 \\ \hline 9 \end{array} \leftarrow 3 \times 3$$

$\boxed{} \leftarrow 20 \times 3$

$$\begin{array}{r} 2\;3 \\ \times \quad\; 3 \\ \hline 9 \end{array} \longrightarrow \begin{array}{r} 2\;3 \\ \times \quad\; 3 \\ \hline \boxed{\;}\;9 \end{array}$$

$\uparrow$ $\uparrow$

3 ones × 3   2 tens × 3

4 Multiply 342 by 2.

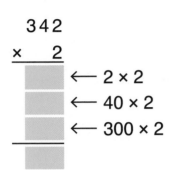

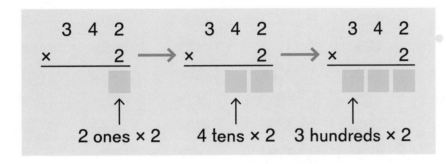

5 Find the value.

(a) 31 × 3   (b) 13 × 3   (c) 2 × 24

(d) 424 × 2   (e) 4 × 221   (f) 202 × 4

6 There are 14 cones in each parking lot.
There are 2 parking lots.
How many cones are there?

Exercise 2 • page 135

Think

The race organizers ordered 42 boxes of cupcakes for a raffle.
Each box has 3 cupcakes.
How many cupcakes are there?

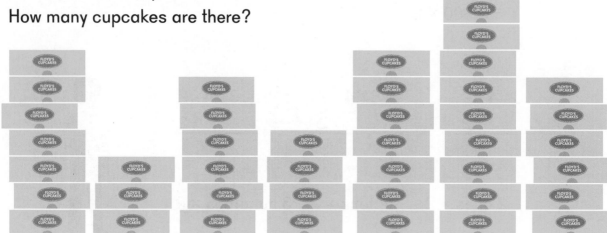

Learn

42 × 3

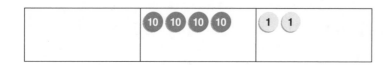

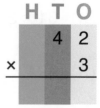

Multiply the ones.

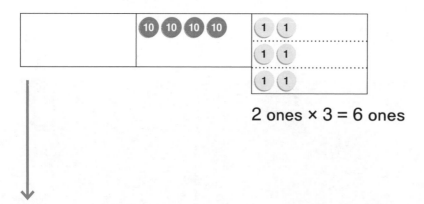

2 ones × 3 = 6 ones

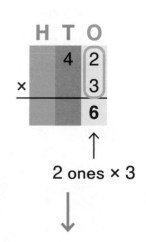

2 ones × 3

Multiply the tens.

Regroup the tens.

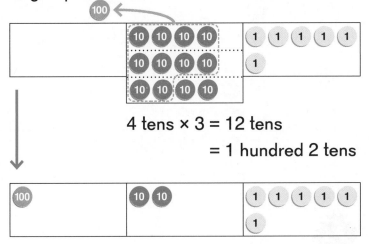

4 tens × 3 = 12 tens

= 1 hundred 2 tens

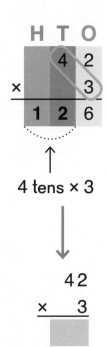

4 tens × 3

$$\begin{array}{r} 42 \\ \times \quad 3 \\ \hline \end{array}$$

$$\begin{array}{r} 42 \\ \times \quad 3 \\ \hline 6 \quad \leftarrow 2 \times 3 \\ 120 \quad \leftarrow 40 \times 3 \\ \hline 126 \end{array}$$

I can use mental math.

42 × 3 = 120 + 6

40   2

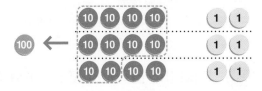

There are ___ cupcakes.

Do

1 Find the product of 83 and 3.

$$\begin{array}{r} 8\ 3 \\ \times\quad\ \ 3 \\ \hline \end{array}$$

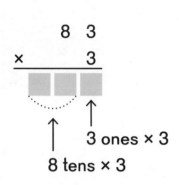

↑ 3 ones × 3

↑ 8 tens × 3

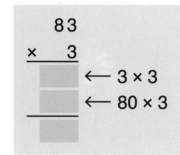

$$\begin{array}{r} 8\,3 \\ \times\quad\ 3 \\ \hline \end{array}$$
← 3 × 3
← 80 × 3

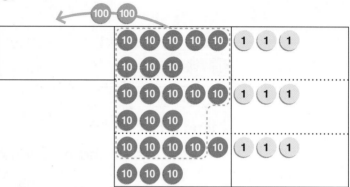

2 Find the product of 2 and 64.

$$\begin{array}{r} 6\ 4 \\ \times\quad\ \ 2 \\ \hline \end{array}$$

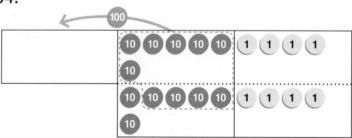

3 Find the product of 51 and 4.

$$\begin{array}{r} 5\ 1 \\ \times\quad\ \ 4 \\ \hline \end{array}$$

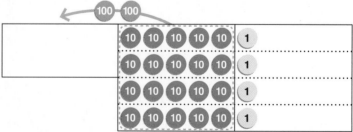

4 Find the product of 3 and 53.

$3 \times 53 = $ ▢

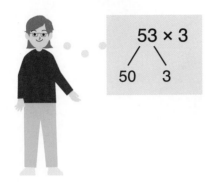

53 × 3
/ \
50 3

5 What are the missing digits?

(a)
```
      7 2
  ×    ▢
  ─────────
    2 8 8
```

(b)
```
      ▢ 1
  ×     5
  ─────────
    3 0 ▢
```

6 Find the value.

(a) 71 × 5

(b) 82 × 4

(c) 42 × 4

(d) 2 × 94

(e) 3 × 93

(f) 81 × 5

7 There are 32 volunteer groups working on the race course.
Each group has 4 members.
How many volunteers are there?

Exercise 3 • page 137

Think

Mei ran 24 miles each week for 3 weeks to prepare for the race.

How many miles did she run to prepare for the race?

Learn

24 × 3

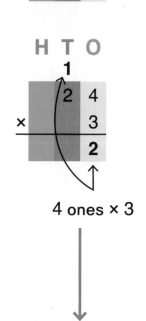

Multiply the ones.
Regroup the ones.

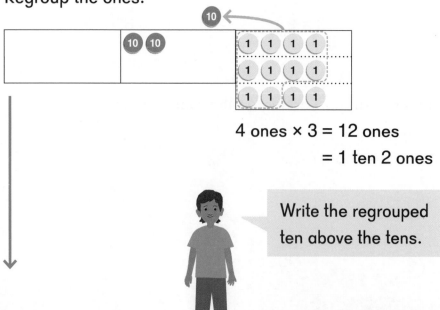

4 ones × 3 = 12 ones

= 1 ten 2 ones

Write the regrouped
ten above the tens.

4 ones × 3

Multiply the tens.

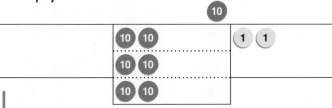

2 tens × 3 = 6 tens

This ten is from multiplying the ones. Do not multiply it again.

Then, add in the regrouped ten.

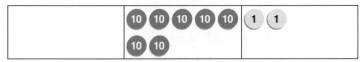

6 tens + 1 ten = 7 tens

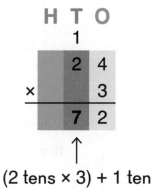

(2 tens × 3) + 1 ten

$$\begin{array}{r} 24 \\ \times\ \ 3 \\ \hline \end{array}$$

```
  24
×  3
─────
  12  ← 4 × 3
  60  ← 20 × 3
─────
  72
```

I can use mental math.
24 × 3 = 60 + 12
20   4

Mei ran ▢ miles.

Do

1 Multiply 14 by 5.

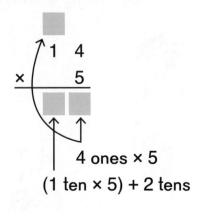

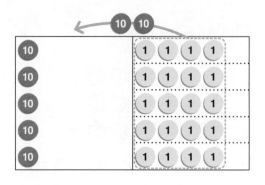

4 ones × 5

(1 ten × 5) + 2 tens

Remember not to multiply the regrouped tens.

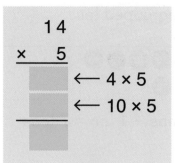

14
× 5

← 4 × 5

← 10 × 5

2 Multiply 3 by 28.

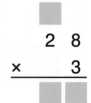

28
× 3

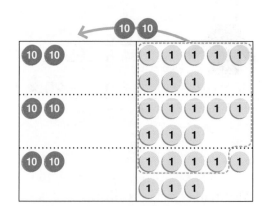

3 Multiply 49 by 2.

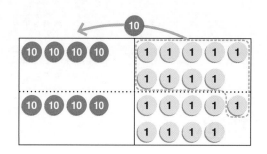

$$
\begin{array}{r}
\square \\
4 \ 9 \\
\times 2 \\
\hline
\square \ \square \\
\end{array}
$$

4 What are the missing digits?

(a)
$$
\begin{array}{r}
\square \\
\square \ 8 \\
\times 4 \\
\hline
7 \ 2 \\
\end{array}
$$

(b)
$$
\begin{array}{r}
\square \\
3 \ 5 \\
\times \square \\
\hline
7 \ 0 \\
\end{array}
$$

5 Find the value.

(a) 17 × 5

(b) 38 × 2

(c) 25 × 3

(d) 24 × 4

(e) 7 × 15

(f) 5 × 19

6

There are 18 sponsors for the race.
Each sponsor donated 3 raffle prizes.
How many raffle prizes are there?

Exercise 4 • page 139

Lesson 5
Multiplication with Regrouping Ones and Tens

Think

One box has 54 registration forms.
A second box has twice as many
forms as the first box.
How many forms are in both boxes?

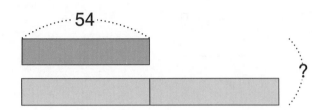

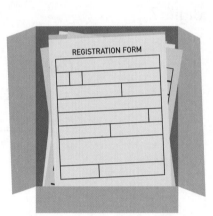

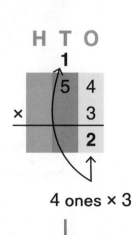

Learn

54 × 3

Multiply the ones.

Regroup the ones.

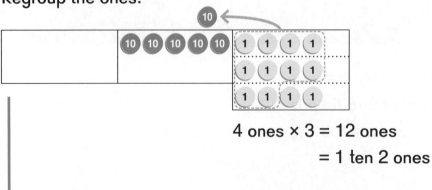

4 ones × 3 = 12 ones
= 1 ten 2 ones

4 ones × 3

Multiply the tens.

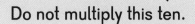

Do not multiply this ten.

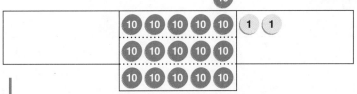

5 tens × 3 = 15 tens

Add in the regrouped ten.
Regroup the total tens.

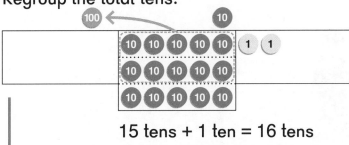

15 tens + 1 ten = 16 tens
16 tens = 1 hundred 6 tens

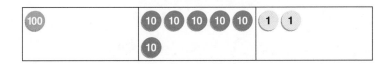

There are ⬜ forms in both boxes.

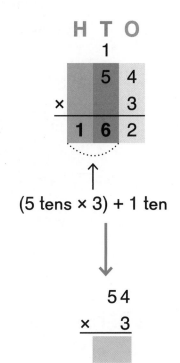

```
    1
    5 4
×     3
─────────
1 6 2
```

(5 tens × 3) + 1 ten

```
    5 4
×     3
─────────
```

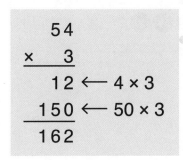

```
      5 4
×       3
──────────
     1 2  ← 4 × 3
   1 5 0  ← 50 × 3
──────────
   1 6 2
```

Do

1 Find the product of 35 and 5.

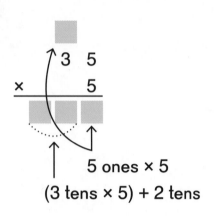

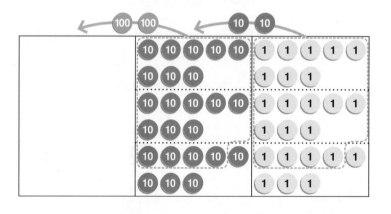

5 ones × 5

(3 tens × 5) + 2 tens

35
× 5

⟵ 5 × 5

⟵ 30 × 5

2 Find the product of 88 and 3.

8 8
× 3

Remember to add in any regrouped tens or hundreds after multiplying.

3 Multiply 79 by 2.

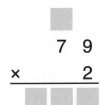

7 9
×   2
▢ ▢ ▢

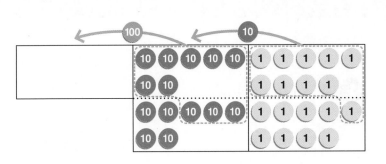

4 What are the missing digits?

(a)
▢
▢ 8
×   4
2 3 ▢

(b)
1
▢ 5
×   ▢
1 3 0

5 Find the value.

(a) 37 × 3

(b) 74 × 5

(c) 52 × 8

(d) 9 × 33

(e) 5 × 62

(f) 7 × 44

6 There are 3 large crates with 50 oranges in each.
There are 3 small crates with 35 oranges in each.
How many oranges are there in all?

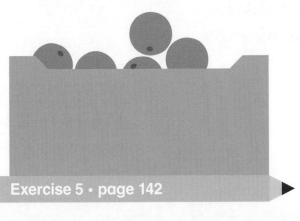

Exercise 5 • page 142

1 Find the value.

(a) 5 × 8   (b) 50 × 8   (c) 500 × 8

(d) 80 × 5   (e) 300 × 9   (f) 4 × 700

(g) 8 × 200   (h) 3 × 900   (i) 800 × 5

2 Find the value.

(a) 23 × 3   (b) 42 × 2   (c) 51 × 5

(d) 8 × 25   (e) 64 × 4   (f) 32 × 8

(g) 45 × 3   (h) 24 × 9   (i) 7 × 43

3 Fliers advertising the race were given to 4 running stores to hand out.
Each store got 400 fliers.
How many fliers were there?

4 One store handed out 40 fliers on the first day.
The next day, it handed out 8 times as many fliers.
If it had 400 fliers to begin with,
how many fliers are left?

**WE WANT YOU
FOR THE DIMENSIONS
MARATHON**

CONTACT MEI TO REGISTER TODAY!

5 Raj made 38 paintings.
He sold each painting for $4.
How much money did he receive?

6 A taco cost $3.
How much do 25 tacos cost?

7 A school cafeteria bought 25 lb of green grapes at $4 a pound and
38 lb of red grapes at $3 a pound.
What was the total cost of the grapes?

8 Jade collected 87 soccer cards.
She collected twice as many basketball cards as soccer cards.
How many cards did she collect altogether?

9 There are 48 wind instruments in the orchestra.
There are 3 times as many string instruments as wind instruments.
How many more string instruments than wind instruments are there?

10 Laila bought 45 vases for $2 each.
After painting them, she sold 38 of them for $5 each.
How much profit did she make?

Exercise 6 • page 145

Think

The race planners ordered 253 packs of medals.
Each pack has 3 medals.
How many medals are there?

Learn

253 × 3

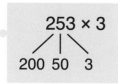

253 × 3
200 50   3

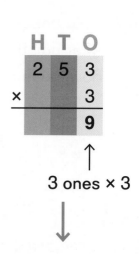

Multiply the ones.

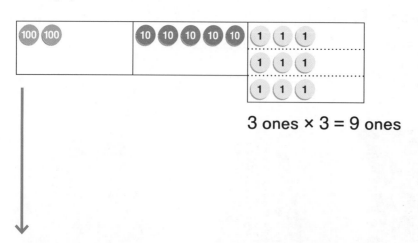

3 ones × 3 = 9 ones

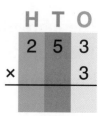

3 ones × 3

Multiply the tens.
Regroup the tens.

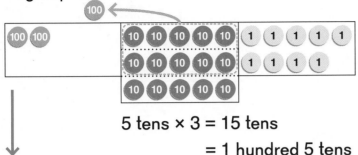

5 tens × 3 = 15 tens
= 1 hundred 5 tens

Multiply the hundreds.

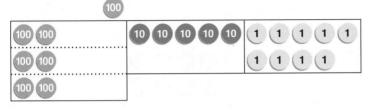

2 hundreds × 3 = 6 hundreds

Add in the regrouped hundred.

6 hundreds + 1 hundred = 7 hundreds

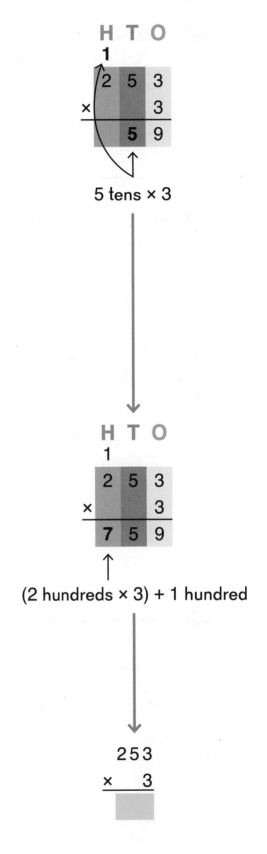

HTO
1
253
×   3
─────
      59

5 tens × 3

HTO
1
253
×   3
─────
7 5 9

(2 hundreds × 3) + 1 hundred

253
×   3
─────

```
    253
×     3
─────
      9  ← 3 × 3
    150  ← 50 × 3
    600  ← 200 × 3
─────
    759
```

There are [   ] medals.

Do

1 Find the product of 432 and 2.

$$\begin{array}{r} 4\ 3\ 2 \\ \times \quad\quad 2 \\ \hline \square\ \square\ \square \end{array}$$

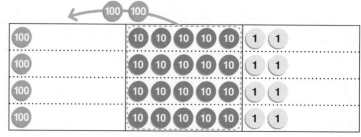

2 Find the product of 152 and 4.

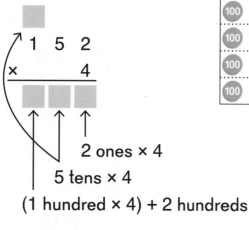

$$\begin{array}{r} \square \\ 1\ 5\ 2 \\ \times \quad\quad 4 \\ \hline \square\ \square\ \square \end{array}$$

2 ones × 4

5 tens × 4

(1 hundred × 4) + 2 hundreds

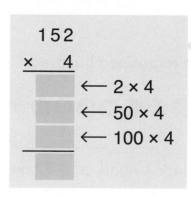

$$\begin{array}{r} 1\ 5\ 2 \\ \times \quad\quad 4 \\ \hline \end{array}$$

← 2 × 4

← 50 × 4

← 100 × 4

3 Find the product of 308 and 3.

$$\begin{array}{r} \square \\ 3\ 0\ 8 \\ \times \quad\quad 3 \\ \hline \square\ \square\ \square \end{array}$$

4 Find the value.

(a) 224 × 3

(b) 162 × 3

(c) 118 × 5

(d) 3 × 209

(e) 5 × 201

(f) 5 × 106

(g) 823 × 2

(h) 3 × 903

(i) 4 × 170

5 Mrs. Choi bought a table and 4 chairs for $1,092.
Each chair cost $116.
How much did the table cost?

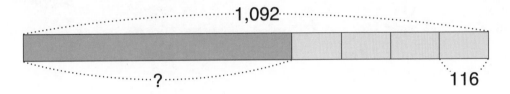

116 × 4 = ⬚ (4 chairs)

1,092 − ⬚ = ⬚ (table)

The table cost $⬚.

6 There were 160 bottles of water in each cooler at the finish line.
There are 5 coolers.
So far, 240 bottles have been taken.
How many bottles remain?

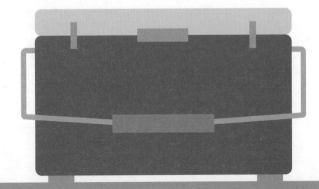

Exercise 7 • page 148

Think

The race planners ordered 345 bottles.
Each bottle cost $3.
How much did all the bottles cost?

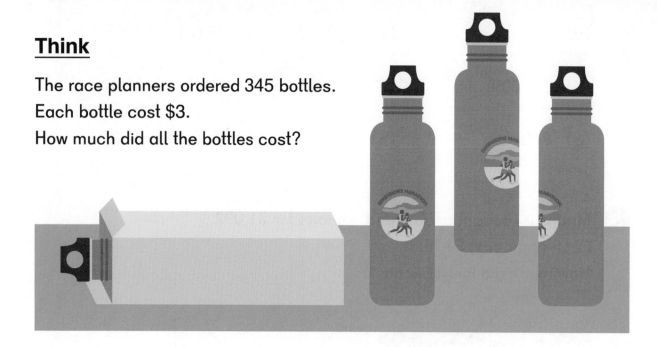

Learn

345 × 3

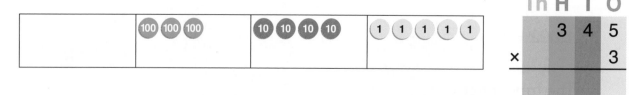

Multiply the ones.
Regroup the ones.

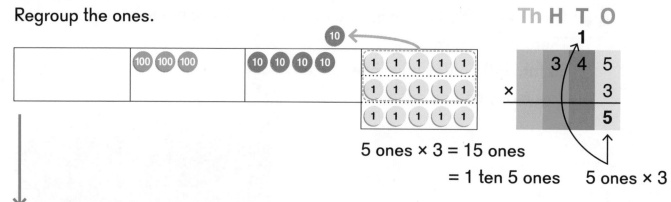

5 ones × 3 = 15 ones
= 1 ten 5 ones

5 ones × 3

Multiply the tens, then add the regrouped ten.
Regroup the total tens.

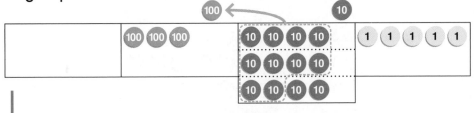

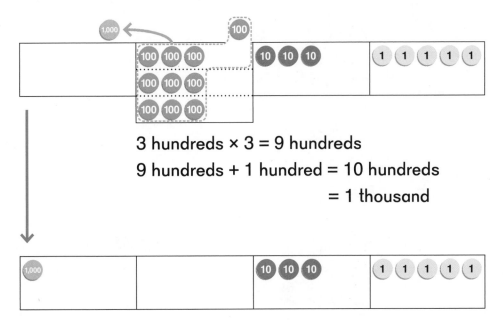

4 tens × 3 = 12 tens

12 tens + 1 ten = 13 tens

= 1 hundred 3 tens

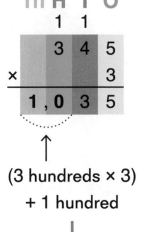

(4 tens × 3) + 1 ten

Multiply the hundreds, then add the regrouped hundred.
Regroup the total hundreds.

3 hundreds × 3 = 9 hundreds

9 hundreds + 1 hundred = 10 hundreds

= 1 thousand

(3 hundreds × 3)
+ 1 hundred

$$\begin{array}{r} 345 \\ \times \quad 3 \\ \hline \end{array}$$

The bottles cost $ _____ .

$$\begin{array}{r} 345 \\ \times \quad 3 \\ \hline 15 \leftarrow 5 \times 3 \\ 120 \leftarrow 40 \times 3 \\ 900 \leftarrow 300 \times 3 \\ \hline 1,035 \end{array}$$

Do

1 Find the product of 867 and 3.

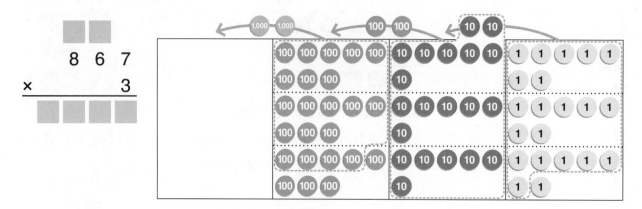

$$\begin{array}{r} 8\ 6\ 7 \\ \times\quad\ 3 \\ \hline \end{array}$$

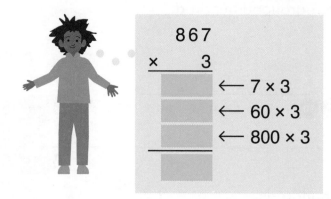

$$\begin{array}{r} 8\ 6\ 7 \\ \times\quad\ 3 \\ \hline \end{array}$$
     ← 7 × 3
     ← 60 × 3
     ← 800 × 3

2 Find the product of 5 and 405.

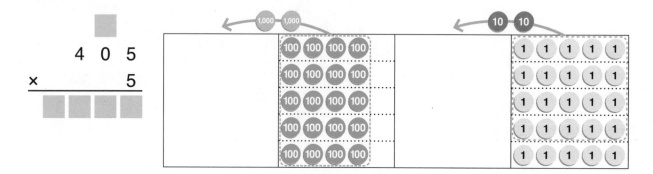

$$\begin{array}{r} 4\ 0\ 5 \\ \times\quad\ 5 \\ \hline \end{array}$$

3 Find the product of 999 and 2.

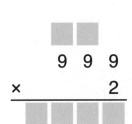

$$\begin{array}{r} 9\,9\,9 \\ \times \qquad 2 \\ \hline \end{array}$$

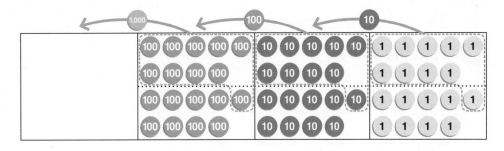

4 Find the product.

(a) 523 × 4

(b) 490 × 5

(c) 637 × 3

(d) 5 × 428

(e) 4 × 706

(f) 2 × 955

5 In a fundraising drive, Gina raised $650.
Tien raised 3 times as much as Gina.
Olga raised $175 less than Tien.
How much money did Olga raise?

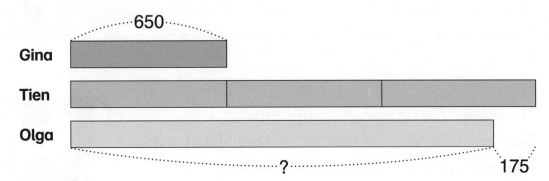

Exercise 8 • page 151

1 Find the value.

(a) 124 × 2

(b) 241 × 3

(c) 4 × 501

(d) 623 × 2

(e) 809 × 5

(f) 760 × 3

(g) 2 × 819

(h) 186 × 3

(i) 3 × 745

(j) 958 × 4

(k) 3 × 759

(l) 354 × 6

2 A mason uses 750 bricks to make a fireplace.
He wants to build 5 fireplaces.
How many bricks does he need?

3 A farm stand sold 115 peaches last week.
This week it sold 5 times as many as last week.
How many peaches did it sell in the two weeks?

4 There are 275 dog biscuits in one box.
Tyler bought 4 boxes of biscuits and fed 18 to his dog.
How many dog biscuits does he have left?

5 A sports store bought 250 baseballs at $3 each.
Then it sold them for $5 each.
How much profit did the store make?

SINCE 1840

DOGGIE 'SCITS

"WOOF!"
-SPOT

HEART SHAPED

Exercise 9 • page 154

Chapter 6

Division

Find the quotients.
What patterns do you see?

2 ÷ 2	3 ÷ 3	4 ÷ 4	5 ÷ 5	10 ÷ 10
4 ÷ 2	6 ÷ 3	8 ÷ 4	10 ÷ 5	20 ÷ 10
6 ÷ 2	9 ÷ 3	12 ÷ 4	15 ÷ 5	30 ÷ 10
8 ÷ 2	12 ÷ 3	16 ÷ 4	20 ÷ 5	40 ÷ 10
10 ÷ 2	15 ÷ 3	20 ÷ 4	25 ÷ 5	50 ÷ 10
12 ÷ 2	18 ÷ 3	24 ÷ 4	30 ÷ 5	60 ÷ 10
14 ÷ 2	21 ÷ 3	28 ÷ 4	35 ÷ 5	70 ÷ 10
16 ÷ 2	24 ÷ 3	32 ÷ 4	40 ÷ 5	80 ÷ 10
18 ÷ 2	27 ÷ 3	36 ÷ 4	45 ÷ 5	90 ÷ 10
20 ÷ 2	30 ÷ 3	40 ÷ 4	50 ÷ 5	100 ÷ 10

Make flash cards for the facts you need to practice.

24 ÷ 4	6
front	back

Think

Some students are sorting donated items equally into 3 containers.
There are 6 boxes of diapers,
60 packages of soap, and 600 individual hand wipes.
How many of each item will go into each container?

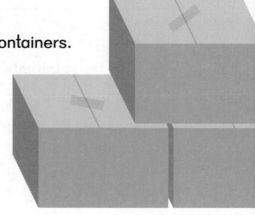

Learn

Divide 6 ones by 3.

     | 6 ones ÷ 3 = 2 ones   **6 ÷ 3 = 2** | $\dfrac{\quad 2}{3\overline{)6}}$

Each container gets [ ] boxes of diapers.

Divide 6 tens by 3.

     | 6 tens ÷ 3 = 2 tens | $\dfrac{\quad 20}{3\overline{)60}}$

Each container gets [ ] packages of soap.

$3 \times \ ? = 60$

Divide 6 hundreds by 3.

     | 6 hundreds ÷ 3 = 2 hundreds | $\dfrac{\quad 200}{3\overline{)600}}$

Each container gets [ ] hand wipes.

Do

1 Divide 60 by 2.

6 tens ÷ 2 = [ ] tens $\qquad$ 60 ÷ 2 = [ ] $\qquad$ 2)60

2 Divide 800 by 4.

8 hundreds ÷ 4 = [ ] hundreds $\qquad$ 800 ÷ 4 = [ ] $\qquad$ 4)800

3 (a) 3 × [ ] = 9 $\qquad$ (b) 3 × [ ] = 90 $\qquad$ (c) 3 × [ ] = 900

9 ÷ 3 = [ ] $\qquad$ 90 ÷ 3 = [ ] $\qquad$ 900 ÷ 3 = [ ]

4 Find the value.

(a) 20 ÷ 2 $\qquad$ (b) 80 ÷ 4 $\qquad$ (c) 50 ÷ 10

(d) 800 ÷ 2 $\qquad$ (e) 500 ÷ 5 $\qquad$ (f) 700 ÷ 10

5 There are 40 fruit snacks.
Each bag will get 2 fruit snacks.
How many bags are needed?

Exercise 1 • page 157

Think

Sofia has 9 bottles of shampoo, 26 hairbrushes, and 43 combs.

She is putting them equally into 2 tubs.

How many of each item will go in each tub?

Will there be any left over?

Learn

Divide 9 bottles of shampoo by 2:

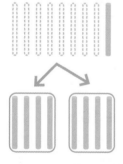

$$\begin{array}{r} 4 \\ 2\overline{)9} \\ 8 \\ \hline 1 \end{array}$$

4 ← 4 ones in each group

8 ← 4 ones × 2

1 ← 9 − 8 (remainder)

9 ÷ 2 is 4 with a remainder of 1.

Check: 9 = 2 × 4 + 1

The quotient is 4.
There is a remainder.

Each tub gets ⬜ bottles of shampoo.

There is ⬜ bottle left over.

Divide 26 hairbrushes by 2:

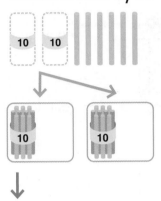

↓

Divide 2 tens by 2.

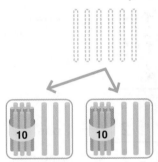

↓

Divide 6 ones by 2.

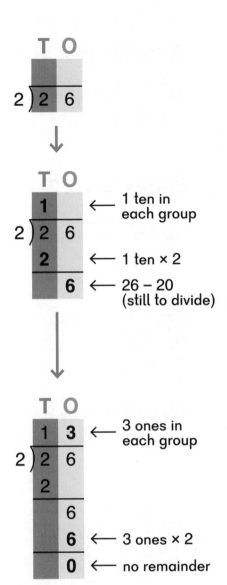

T O

2)2 6

↓

1 ten in each group ←

1 ten × 2 ←

26 − 20 (still to divide) ←

↓

3 ones in each group ←

3 ones × 2 ←

no remainder ←

26 ÷ 2
  /   \
20   6

26 ÷ 2 = 13

Check: 13 × 2 = 26

Each tub gets [ ] hairbrushes.

The quotient is 13.
There is no remainder.
We can say the
remainder is 0.

Divide 43 combs by 2:

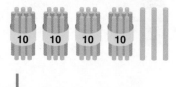

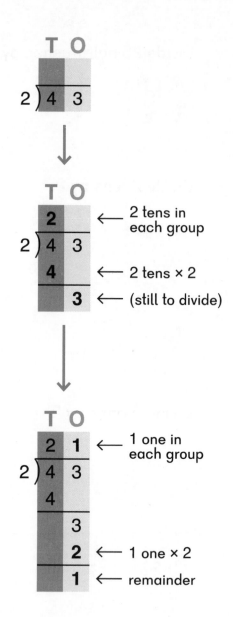

Divide 4 tens by 2.

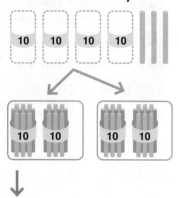

2 tens in
each group

2 tens × 2

(still to divide)

Divide 3 ones by 2.

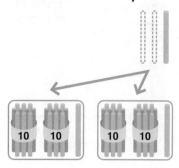

1 one in
each group

1 one × 2

remainder

43 ÷ 2 is 21 with a remainder of 1.

Check: 21 × 2 + 1 = 43

Each tub gets ▢ combs.

There is ▢ comb left over.

43 ÷ 2

40   3

Do

1 Divide 68 by 2.

 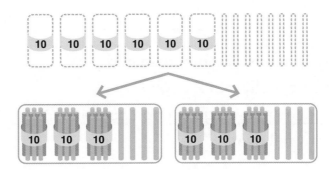

2 Divide 27 by 2.

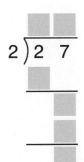 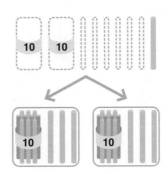

3 Find the quotient and remainder.

(a) $49 \div 2$ (b) $28 \div 2$ (c) $46 \div 2$

(d) $85 \div 2$ (e) $67 \div 2$ (f) $88 \div 2$

4 Mei is putting the 21 combs from one of the tubs into bags.
Each bag will have 2 combs.
How many bags does she need?
Will there be any combs left over?

Exercise 2 • page 159

Think

Alex is putting 14 tubes of toothpaste and
53 toothbrushes equally into 2 bags.
How many of each item will go in each container?
Will there be any left over?

Learn

Divide 14 tubes of toothpaste by 2:

↓

1 ten 4 ones = 14 ones

↓

Divide 14 ones by 2.

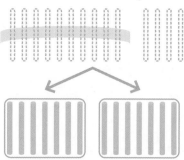

```
  T O
2 ) 1 4
```

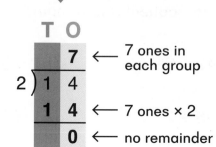

```
    T O
      7   ← 7 ones in
2 ) 1 4       each group
    1 4   ← 7 ones × 2
      0   ← no remainder
```

$14 \div 2 = 7$ | Check: $7 \times 2 = 14$

Each bag gets _____ tubes of toothpaste.

Divide 53 toothbrushes by 2:

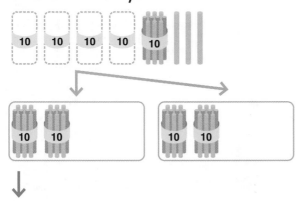

↓

Divide 5 tens by 2.

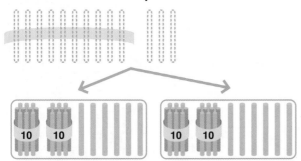

↓

Regroup the remaining tens.
Divide 13 ones by 2.

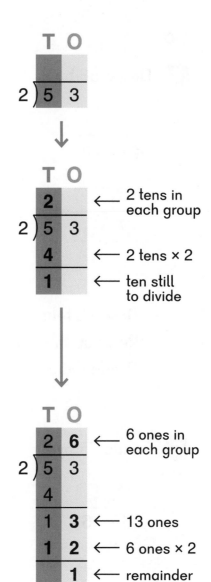

2 tens in each group

2 tens × 2

ten still to divide

6 ones in each group

13 ones

6 ones × 2

remainder

53 ÷ 2 is 26 with a remainder of 1. │ Check: 26 × 2 + 1 = 53

Each bag gets ▢ toothbrushes.

There is ▢ toothbrush left over.

Do

1 Divide 56 by 2.

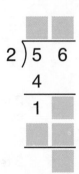

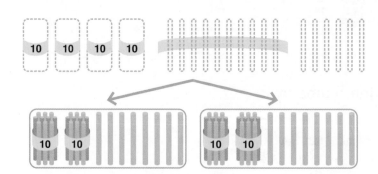

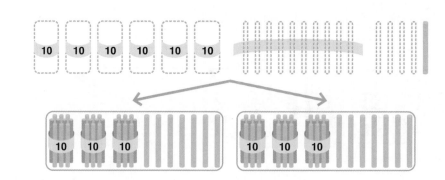

Divide 5 tens by 2.
Regroup the remaining tens.
Divide the ones by 2.

2 Divide 75 by 2.

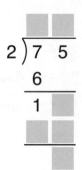

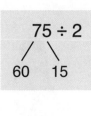

$$75 \div 2$$

60   15

3 Divide 91 by 2.

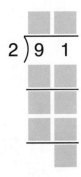

4 Find the quotient and remainder.

(a) 33 ÷ 2   (b) 72 ÷ 2   (c) 55 ÷ 2

(d) 77 ÷ 2   (e) 96 ÷ 2   (f) 38 ÷ 2

5 Which of the following numbers will have a remainder if divided by 2?
Are they even numbers or odd numbers?

(a) 9   (b) 14   (c) 26   (d) 43

(e) 56   (f) 75   (g) 91   (h) 100

 When an even number is divided by 2, the remainder is 0.
The ones digit of even numbers is 0, 2, 4, 6, or 8.
When an odd number is divided by 2, the remainder is 1.
The ones digit of odd numbers are 1, 3, 5, 7, or 9.

Exercise 3 • page 161

Think

(a) Dion is dividing 76 bars of soap equally into 3 tote bags.
How many bars of soap will each tote bag get?
How many bars of soap are left over?

(b) Mei has 76 granola bars.
She is putting 3 granola bars in each sandwich bag.
How many sandwich bags does she need?
How many granola bars are left over?

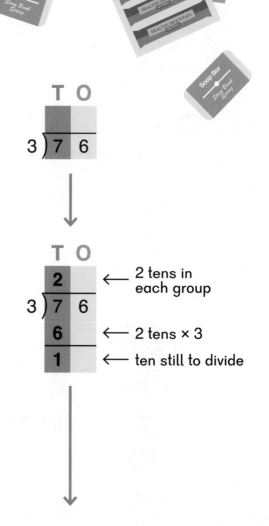

Learn

(a) 76 ÷ 3 = ?

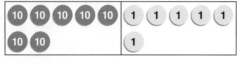

Divide 7 tens by 3.

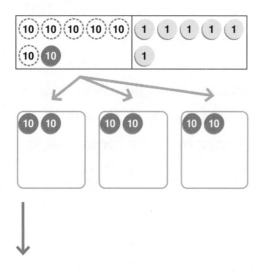

T O

3)7 6

2 tens in each group ←

3)7 6

6 ← 2 tens × 3

1 ← ten still to divide

Regroup the remaining tens.
Divide 16 ones by 3.

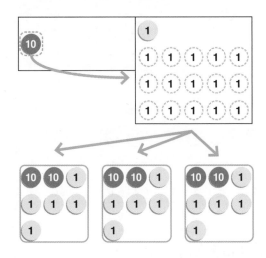

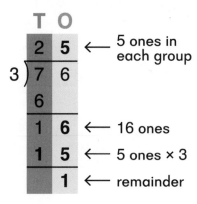

5 ones in each group

← 16 ones

← 5 ones × 3

← remainder

76 ÷ 3 is 25 with a remainder of 1. │ Check: 25 × 3 + 1 = 76

Each tote bag gets [   ] bars of soap and there is [   ] bar of soap left over.

(b)

```
    2 5
 3)7 6
    6        ← For 60 granola bars, we need 20 bags.
   ───
    1 6
    1 5      ← For 16 granola bars, we need 5 bags.
   ───
      1      ← 1 granola bar is left over.
```

76 ÷ 3
  ╱ ╲
60   16

Even though the situation involves grouping, we can still use the same method.

Mei needs [   ] sandwich bags.

There is [   ] granola bar left over.

Do

 Divide 98 by 4.

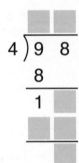

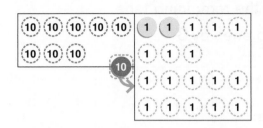

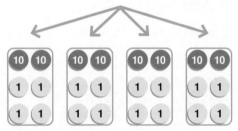

The remainder should be less than...

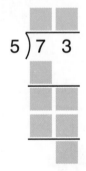 Divide 73 by 5.

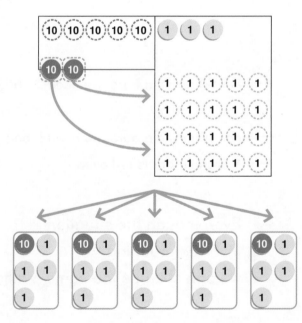

3 Divide 84 by 4.

$$84 \div 4 = \boxed{}$$

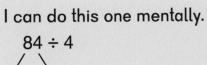
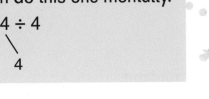

I can do this one mentally.

$$84 \div 4$$

80   4

4 Find the quotient and remainder.

(a) $68 \div 4$ (b) $72 \div 3$ (c) $90 \div 5$

(d) $53 \div 3$ (e) $87 \div 4$ (f) $83 \div 5$

(g) $37 \div 5$ (h) $98 \div 3$ (i) $53 \div 4$

5 What are the missing digits?

(a)
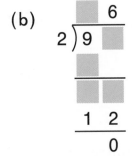

(b)

6 A 76-ft long rope is cut into 4-ft long pieces.
How many pieces of rope are there?

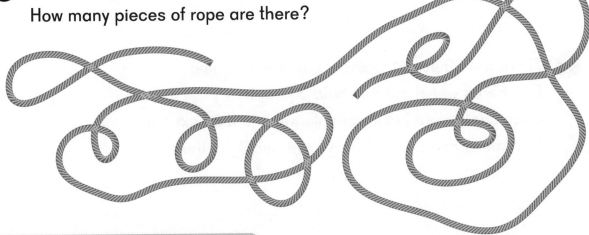

Exercise 4 • page 164

1 Find the quotient.

(a) 600 ÷ 2

(b) 90 ÷ 3

(c) 800 ÷ 4

(d) 80 ÷ 2

(e) 500 ÷ 5

(f) 200 ÷ 2

2 Find the quotient and remainder.

(a) 46 ÷ 2

(b) 81 ÷ 3

(c) 79 ÷ 4

(d) 65 ÷ 5

(e) 88 ÷ 3

(f) 69 ÷ 2

(g) 72 ÷ 4

(h) 27 ÷ 2

(i) 99 ÷ 5

(j) 77 ÷ 3

(k) 93 ÷ 5

(l) 83 ÷ 2

3 Wainani is putting 75 rolls of paper towels equally into 4 boxes.
How many rolls of paper towels will be in each box?
How many will be left over?

4 There are 89 people going on the Ferris wheel.
One Ferris wheel car can hold 3 people.
Each car is filled before people get into the next car.
How many Ferris wheel cars have people in them?

5 Hunter has 75 flashlights to put into boxes.
Each box will get 3 flashlights.
He first tested each and found that 6 did not work and discarded them.
How many boxes did he use?

6 Aurora saved $40 last month and $55 this month.
She wants to buy gifts that cost $5 each for her friends.
How many gifts can she buy?

7 The sum of two numbers is 100.
The difference between the two numbers is 28.
What are the two numbers?

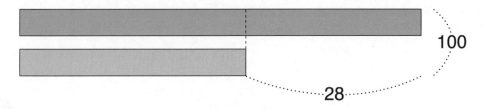

100

28

8 Adam, Chapa, and Ethan saved a total of $92.
Chapa saved $7 more than Adam.
Ethan saved $12 more than Chapa.

If I make Adam's bar
1 unit, then I can
subtract 7, 12, and 7
to have 3 equal units.

(a) How much money did Adam save?

Adam

7

Chapa

12

Ethan

92

(b) How much money did Chapa save?

(c) How much money did Ethan save?

Exercise 5 • page 167

Think

Emma is putting 531 scarves equally into 2 containers.

How many will go in each container?

How many scarves are left over?

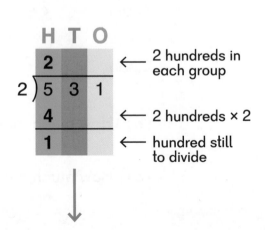

Learn

531 ÷ 2

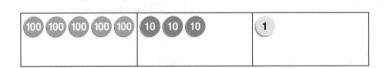

I know 531 is an odd number so there will be a remainder.

$$2 \overline{) \begin{array}{ccc} 5 & 3 & 1 \end{array}}$$

Divide 5 hundreds by 2.

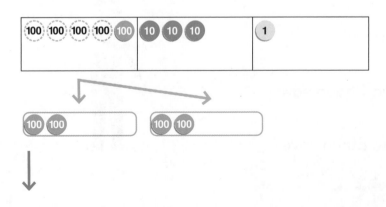

← 2 hundreds in each group

← 2 hundreds × 2

← hundred still to divide

Regroup the remaining hundreds,
then divide 13 tens by 2.

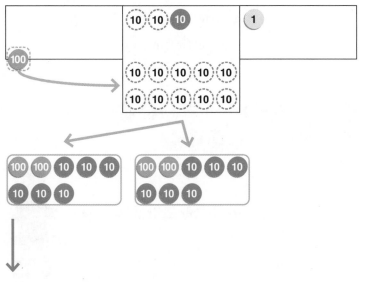

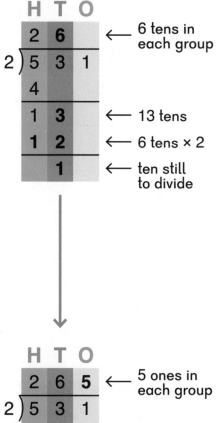

← 6 tens in
each group

← 13 tens

← 6 tens × 2

← ten still
to divide

Regroup the remaining tens,
then divide 11 ones by 2.

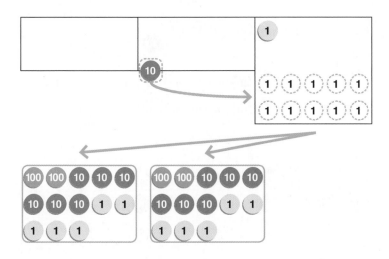

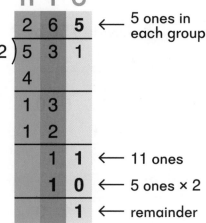

← 5 ones in
each group

← 11 ones

← 5 ones × 2

← remainder

531 ÷ 2 is 265 with a remainder of 1. │ Check: 265 × 2 + 1 = 531

Each container gets ▢ scarves.

There is ▢ scarf left over.

Do

1 Divide 700 by 2.

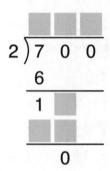

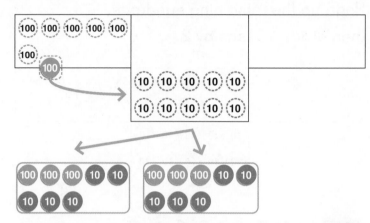

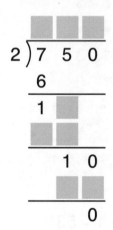

Divide 7 hundreds by 2 first.

2 Divide 750 by 2.

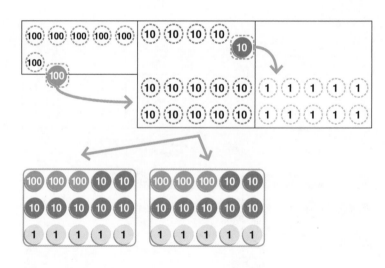

Divide hundreds,
then tens,
then ones.

3 Divide 935 by 2.

Divide the hundreds. → Divide the tens. → Divide the ones.

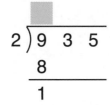

$$2\overline{)9\ 3\ 5}$$
$$\quad 8$$
$$\quad \overline{1}$$

$$2\overline{)9\ 3\ 5}$$
$$\quad 8$$
$$\quad \overline{1}$$

$$2\overline{)9\ 3\ 5}$$
$$\quad 8$$
$$\quad \overline{1}$$

1

1

4 Divide 864 by 2.

$864 \div 2 = \boxed{}$

This one is easy to solve mentally.

$$864 \div 2$$
$$\diagup\ \ |\ \ \diagdown$$
$$800\ \ 60\ \ \ 4$$

5 Find the quotient and remainder.

(a) $870 \div 2$ (b) $475 \div 2$ (c) $904 \div 2$

(d) $501 \div 2$ (e) $878 \div 2$ (f) $999 \div 2$

6 2 identical bicycles cost $504.
How much does 1 such bicycle cost?

Exercise 6 • page 170

Think

Sofia is putting 470 hats equally into 3 containers.
How many hats will go in each container?
How many hats are left over?

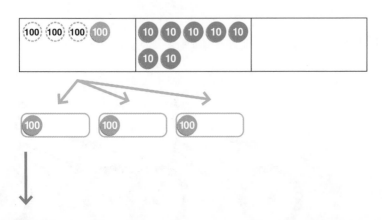

Learn

The remainder must be less than 3...

470 ÷ 3

Divide 4 hundreds by 3.

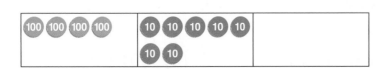

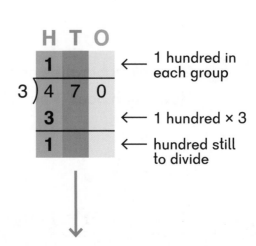

← 1 hundred in each group

← 1 hundred × 3

← hundred still to divide

Regroup the remaining hundreds,
then divide 17 tens by 3.

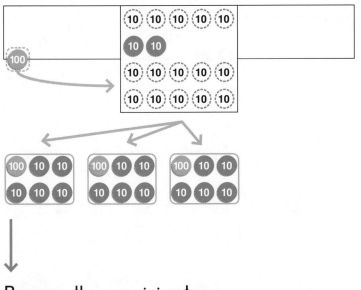

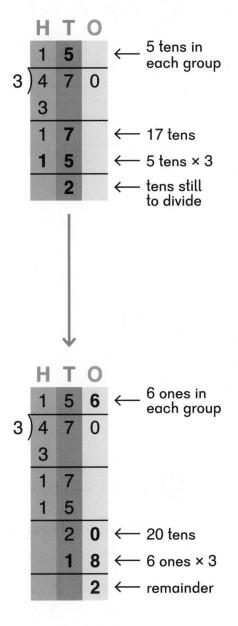

5 tens in
each group

← 17 tens

← 5 tens × 3

← tens still
to divide

Regroup the remaining tens,
then divide 20 ones by 3.

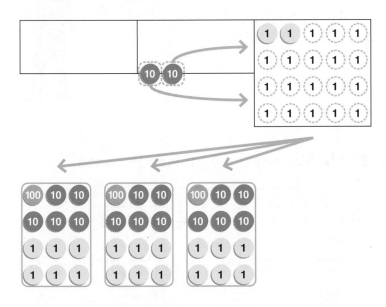

6 ones in
each group

← 20 tens

← 6 ones × 3

← remainder

470 ÷ 3 is 156 with a remainder of 2. │ Check: 156 × 3 + 2 = 470

Each container gets ⬚⬚⬚ hats.

There are ⬚⬚⬚ hats left over.

Do

1 Divide 523 by 5.

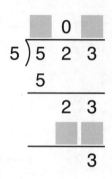

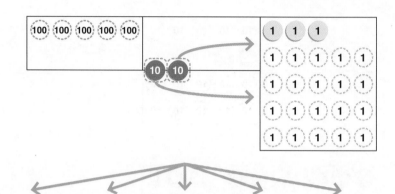

Divide the hundreds.
There are not enough tens to
divide, so regroup all the tens.

2 Divide 943 by 4.

Divide the hundreds. → Divide the tens. → Divide the ones.

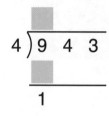

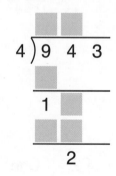

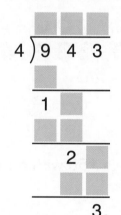

3 Divide 809 by 3.

Divide the hundreds. → Divide the tens. → Divide the ones.

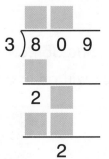

4 Divide 639 by 3.

$639 \div 3 = $

This one is easy to solve mentally.

$639 \div 3$

600 30 9

5 Find the quotient and remainder.

(a) $510 \div 5$ (b) $460 \div 4$ (c) $724 \div 5$

(d) $842 \div 3$ (e) $505 \div 4$ (f) $630 \div 5$

6 3 identical surfboards cost $834.
How much does 1 such surfboard cost?

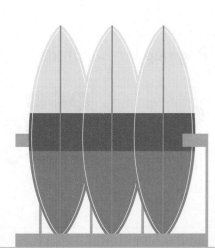

Exercise 7 • page 174

Think

Mei is putting 130 granola bars and 364 cookies equally into 4 containers.
How many of each kind of snack will go in each container?
Will there be any left over?

Learn

Divide 130 by 4.

$130 \div 4$

I can't divide 1 hundred into 4 groups,
so I will regroup it as 10 tens.
The quotient will be a 2-digit number.

Regroup the hundreds.

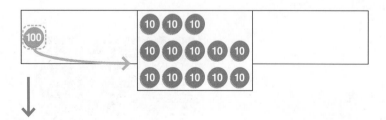

Divide the tens.

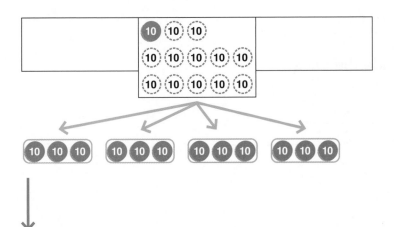

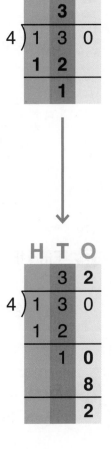

Regroup the remaining tens, then divide the ones.

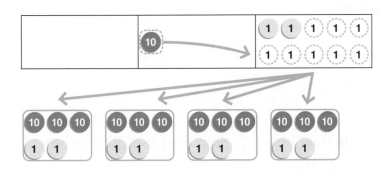

130 ÷ 4 is 32 with a remainder of 2.

Each container gets [ ] granola bars.

There are [ ] granola bars left over.

Divide 364 by 4.

364 ÷ 4 = [ ]

Each container gets [ ] cookies.

There are [ ] cookies left over.

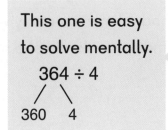

This one is easy
to solve mentally.

364 ÷ 4

360   4

Do

1 Divide 485 by 5.

Divide the tens. → Divide the ones.

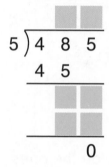

2 Find the quotient and remainder.

(a) 278 ÷ 3   (b) 278 ÷ 4   (c) 278 ÷ 5

3 First, decide whether the quotient will be a 2-digit or a 3-digit number. Then, find the quotient and remainder.

(a) 175 ÷ 2   (b) 560 ÷ 4   (c) 224 ÷ 5

(d) 242 ÷ 3   (e) 305 ÷ 2   (f) 520 ÷ 3

4 Onowa collected 340 washcloths as donations. She had 4 times as many washcloths as towels. How many towels did she have?

Exercise 8 • page 177

1 Find the quotient.
Try to solve mentally.

(a) $64 \div 2$

(b) $903 \div 3$

(c) $364 \div 4$

(d) $80 \div 5$

(e) $824 \div 4$

(f) $255 \div 5$

2 Find the quotient and remainder.

(a) $900 \div 2$

(b) $809 \div 2$

(c) $297 \div 3$

(d) $192 \div 4$

(e) $242 \div 3$

(f) $197 \div 2$

(g) $345 \div 4$

(h) $787 \div 3$

(i) $167 \div 3$

(j) $291 \div 4$

(k) $459 \div 5$

(l) $409 \div 5$

3 A baseball coach has $187.
He wants to buy baseballs that cost $5 each.
How many baseballs can he buy?

4 Jason has 210 m of ribbon.
He wants to cut it into pieces each 4 m long.
How many pieces will he have?

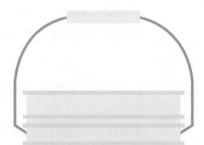

Baseballs

5 A store owner put 225 apples into bags
of 5 apples each.
She sold the bags for $7 each.
How much money did she receive?

Orchard Fresh Fruit

6 Austin saved $165.
He saved 3 times as much money as Kona.
How much more money did he save than Kona?

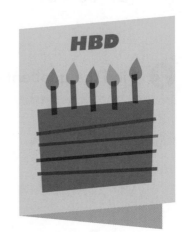

7 Tatiana made 128 cards each week for 3 weeks.
Then, she put them equally into 4 boxes.
How many cards are in each box?

8 The sum of two numbers is 560.
The difference between them is 90.
What are the two numbers?

If I make Sara's bar
1 unit, then I can
add 20 to have 5
equal units.

9 Anna, Pablo, and Sara saved a total of $155.
Anna saved twice as much as Sara.
Pablo saved $20 less than Anna.

(a) How much money did Sara save?

Sara
Anna   155
Pablo
                                          20

(b) How much money did Anna save?

(c) How much money did Pablo save?

Exercise 9 • page 180

Chapter 7

Graphs and Tables

Think

Emma **surveyed** campers to find out which camp activity they liked the best. She recorded her **data** with tally marks.

Activity	Tally	Number
Nature hikes	卌 卌 卌 卌 卌 ////	
Science talks	卌 卌 卌 卌	
Crafts	卌 卌 卌 卌 //	
Campfire cooking	卌 卌 /	
Water sports	卌 卌 卌 卌 卌 卌 ///	

(a) Display the information in a graph.
     Use one ▇ to show 5 students.

How can we show 1, 2, 3, or 4 students?

(b) Which two activities were liked the most?

(c) How many more campers liked water sports than science talks?

(d) How many fewer campers liked the campfire cooking than crafts?

Learn

(a)

Favorite Activity at Camp				
Nature hikes	Science talks	Crafts	Campfire cooking	Water sports

Each ▮ stands for 5 people.

I used ▬, ▭, ▮, and ▮ for 1, 2, 3, and 4.

I can quickly see on the picture graph which activity was the favorite.

(b) _____ and _____ were liked the most.

(c) ▮ more campers liked water sports than science talks.

(d) ▮ fewer campers liked the campfire cooking than crafts.

This **bar graph** shows the same information.

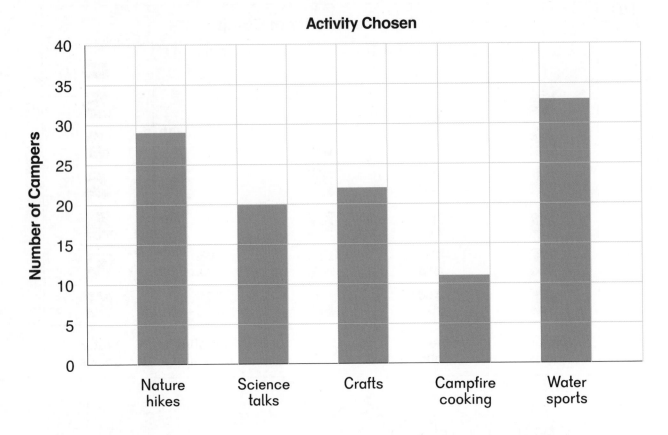

Activity Chosen

The **scale** on the left side is marked in increments of 5 campers.

Bar graphs make it easy to see at a glance which category has more or has less.

List the activities in order from least liked to most liked.

Do

1 This bar graph shows how many badges were awarded at camp for 8 different types of activities.

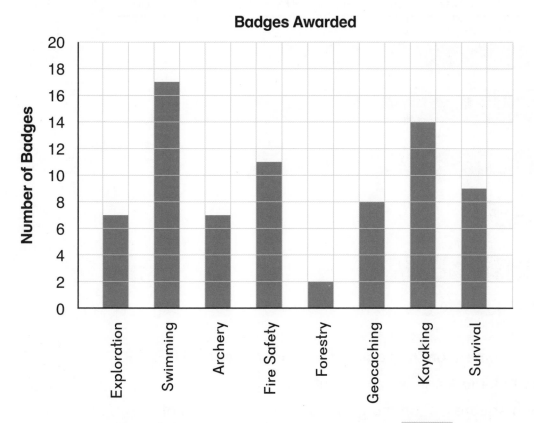

Badges Awarded

(a) The scale is marked in increments of ☐.

(b) Which badge was awarded the most?

(c) Which badge was awarded the least?

(d) How many more Survival badges were awarded than Exploration badges?

(e) How many fewer Archery badges were awarded than Kayaking badges?

(f) How many badges were awarded altogether?

2 This bar graph shows the results of a survey of the favorite hiking routes.

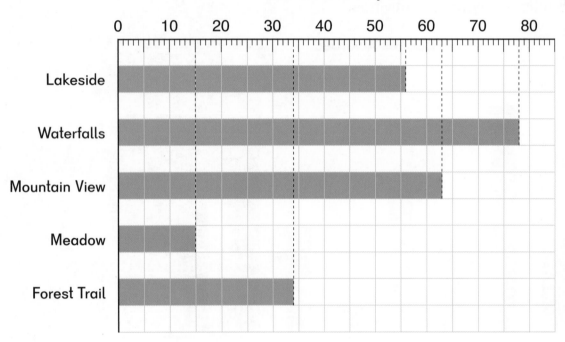

Number of Campers

(a) The numbers on the scale show in increments of ⬜.

(b) The tick marks show increments of ⬜.

(c) Each square ⬜ on the graph shows increments of ⬜.

(d) List the hikes in order from most liked to least liked.

3 For each graph, give the increments shown by the numbers on the scale and by 1 square on the graph, and the quantity shown by each bar.

(a)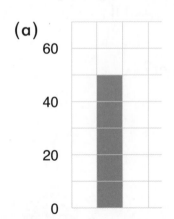

(b)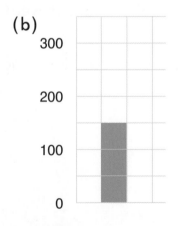

(c)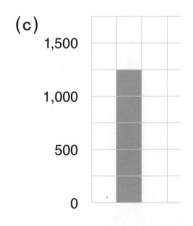

Exercise 1 • page 183

Think

Some campers collected data on how many of 6 different types of trees they found by the lake to see which type grew best in that environment.

Tree	Aspen	Cedar	Elm	Fir	Sycamore	Willow
Number	15	32	70	45	78	92

Use graph paper to draw a bar graph.

① Determine the scale by making sure the largest quantity in the data can fit in the graph.

② Write numbers on the scale and specify the units (number of trees). Determine how wide to make the bars so all of them fit.

③ Write the categories along the bottom.

④ Draw bars corresponding to the data.

⑤ Write a title.

Learn

Tree	Aspen	Cedar	Elm	Fir	Sycamore	Willow
Number	15	32	70	45	78	92

Dion drew this graph.

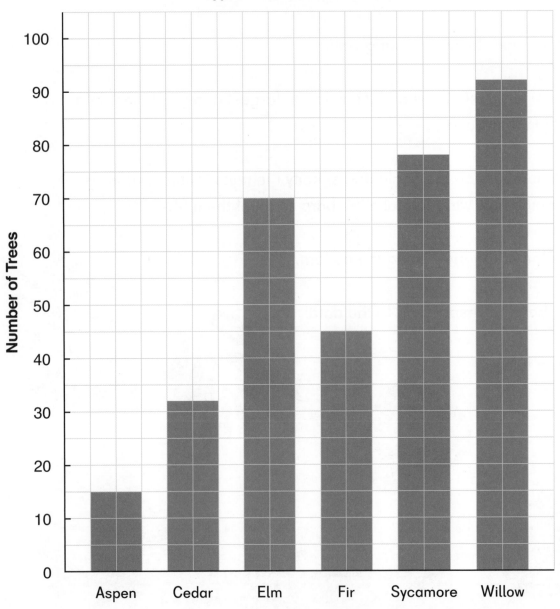

Types of Trees Near the Lake

Do

1 Use the graph or the table to answer the questions.

(a) The scale is marked in increments of ▢.
(b) Each square ▢ on the graph shows increments of ▢.
(c) How many total trees did they count?
(d) Which kind of tree was there the most of?
(e) Which kind of tree was there the least of?
(f) How many fewer elms than sycamores were there?
(g) How many more willows than aspens were there?
(h) Which types of trees grow best in this area?

2 Mei drew this graph.
Why might she have organized the information this way instead?

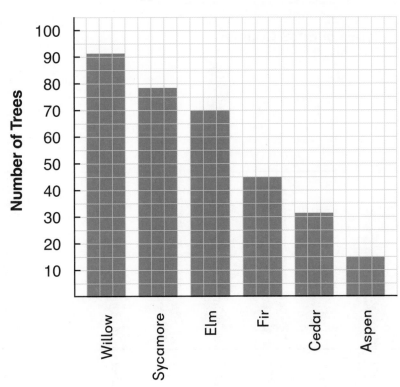

Types of Trees Near the Lake

3 Some scientists gave a presentation about bird migration in some wetlands near the camp. The table and graph show the number of waterfowl counted over a certain period of time.

Bird	Number
Canada goose	1,062
Gadwall	40
Green-winged teal	865
Blue-winged teal	582
Northern shoveler	189
Northern pintail	52
Ring-necked duck	341
Goldeneye	698
Mallard	1,172
Canvasback	31
American coot	692

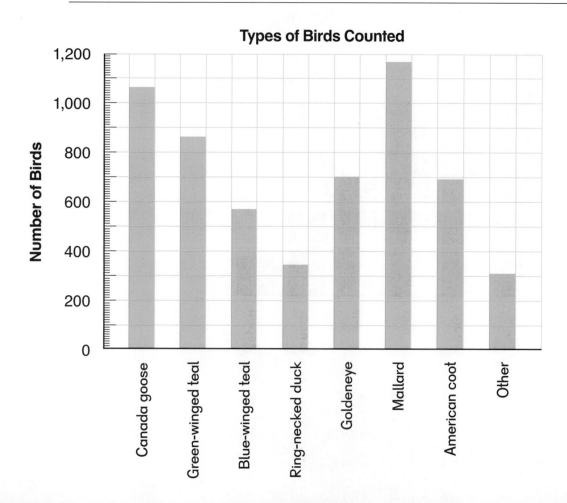

Types of Birds Counted

In the graph, the category "other" was created
to include birds that were in small numbers.
Which birds are shown in this category on the graph?

The category Other is
often put at the end.

4 Answer the following questions for the birds.
Did you use the graph, the table, or both for each?

(a) Approximately how many birds were counted for this data?

(b) Which bird was seen the most?

(c) List the types of birds in order from least to greatest number.

(d) How many teals were seen in all?

(e) For which types were there more than 600 birds counted?

(f) For which types were there less than 600 birds counted?

(g) There were a little over 300 of which type of bird?

5 These graphs were made from data on favorite campfire foods.

Graph 1

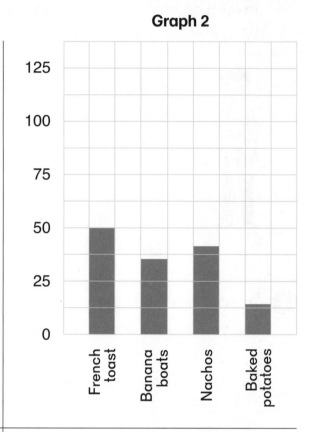

Graph 2

(a) What is the same about each graph?

(b) What is different about each graph?

(c) On which graph is it easiest to tell the actual numbers?

(d) Which graph makes it look like there is not much difference in favorite campfire foods and why?

Graph 3

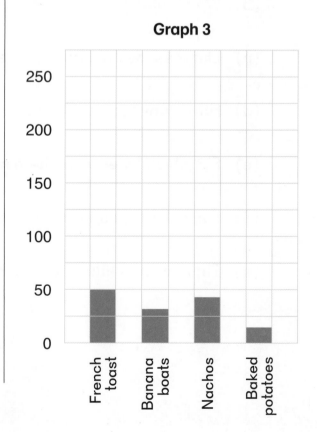

1 This is a survey of some water sports that campers would like to do. Campers could vote for up to two activities.

Activity	Number
Kayak	61
Canoe	35
Sail	59
Snorkel	19
Wakeboard	73

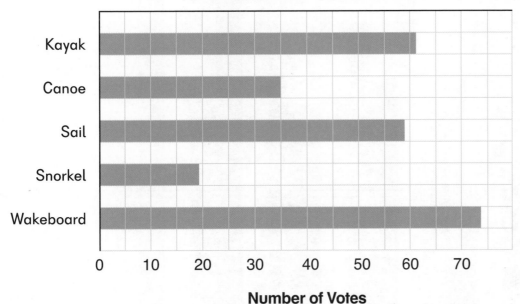

Water Sports Campers Voted For

Number of Votes

(a) What increment is represented by 1 square on the graph?

(b) List the activities in order from least to greatest number of times voted for.

(c) Which two activities were voted for almost the same number of times?

(d) What is the difference in number of votes between the most popular and the least popular activity?

2 Some students wore pedometers and recorded the total number of steps they took each day for two days.
They drew a graph to show the results.

Name	Day 1	Day 2
Wyatt	2,416	3,290
Fang	1,504	4,578
Mariya	4,892	6,009
Noah	4,509	5,605
Carlos	6,953	4,570
Kaylee	6,561	3,350

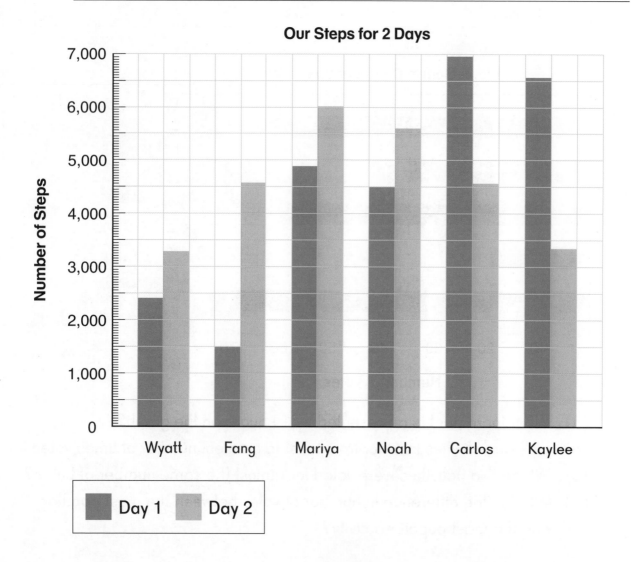

(a) Each square on the graph shows an increment of ▢ .
(b) Who recorded more steps on Day 1 than Day 2?
(c) Who recorded more steps on Day 2 than Day 1?
(d) Who recorded the most steps for both days combined? Use estimation.
(e) Who recorded the fewest steps for both days combined? Use estimation.
(f) Who had the greatest difference in number of steps between the two days?
(g) Who had the least difference in number of steps between the two days?
(h) List the total number of steps each student recorded for both days in order from least to greatest.
(i) List the difference in number of steps each student recorded for both days in order from least to greatest.

3 (a) Use the chart to tally each type of injury at camp.

Bruise	Bug Bite	Burn	Scrape	Bug Bite	Cut	Bruise
Burn	Cut	Rash	Cut	Rash	Scrape	Scrape
Scrape	Scrape	Scrape	Bug Bite	Bug Bite	Sprain	Scrape
Bruise	Bruise	Cut	Bug Bite	Cut	Bug Bite	Cut
Rash	Bug Bite	Burn	Bruise	Bug Bite	Bug Bite	Scrape
Bug Bite	Bruise	Cut	Scrape	Rash	Bruise	Rash
Cut	Rash	Rash	Scrape	Cut	Bruise	Bruise
Bee Sting	Burn	Bug Bite	Sprain	Rash	Scrape	Scrape
Scrape	Scrape	Cut	Burn	Scrape	Bug Bite	Cut
Rash	Cut	Scrape	Scrape	Bruise	Bruise	Cut
Scrape	Scrape	Cut	Bee Sting	Cut	Sprain	Scrape
Cut	Rash	Rash	Bug Bite	Cut	Bruise	Bee Sting
Cut	Burn	Scrape	Scrape	Bug Bite	Cut	Cut
Bug Bite	Cut	Rash	Burn	Bug Bite	Scrape	Scrape

(b) Copy and complete this table from the data.

Injury	Bruise	Cut	Rash	Scrape	Bug Bite	Other
Number						

(c) Draw a bar graph for this data.

(d) Which was the most common injury?

(e) List the five next most common injuries in order of most common to least common.

(f) How might this data help camp counselors?

Exercise 3 • page 192

7-3 Practice

① Estimate, then find the value.

(a) 4,890 + 283

(b) 6,785 + 2,295

(c) 9,125 − 864

(d) 9,006 − 4,237

② Find the value.

(a) 65 × 4

(b) 70 × 0

(c) 98 × 3

(d) 849 × 2

(e) 307 × 9

(f) 474 ÷ 5

(g) 67 ÷ 3

(h) 59 ÷ 4

(i) 963 ÷ 2

(j) 85 ÷ 1

(k) 85 ÷ 85

(l) 0 ÷ 85

③ (a) [   ] ÷ 4 is 9 with a remainder of 3.

(b) [   ] ÷ 3 is 48 with a remainder of 1.

④ (a) What is the least odd number that can be formed using the digits 7, 8, 0, and 3?

(b) What is the greatest even number that can be formed using the digits 6, 3, 5, and 9?

5 This table and graph below shows the number of campsites rented at a park each night for one week.

On Friday, Saturday, and Sunday the cost for renting a campsite is $5 a night.

On Monday through Thursday, the cost for renting a campsite is $4 a night.

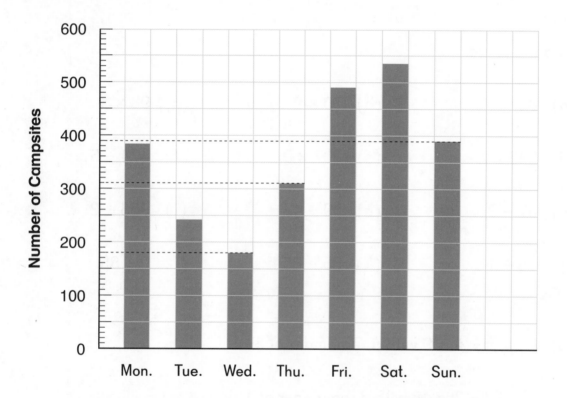

(a) The scale is numbered in increments of ⬚.

(b) Each tick mark shows an increment of ⬚.

(c) Each square on the graph is for an increment of ⬚.

(d) On which day of the week were the greatest number
of campsites rented?

(e) On which day of the week were the fewest number
of campsites rented?

(f) Which two days had about the same number of campsites rented?

(g) Complete the table.

Day	Mon.	Tue.	Wed.	Thu.	Fri.	Sat.	Sun.
Number	382	241			493	532	

(h) How many more campsites were rented on Saturday than on Sunday?

(i) Estimate how many campsites were rented all week by rounding
the daily rental to the nearest 100.

(j) How much money did the camp receive on Friday?

(k) How much money did the camp receive Tuesday through Thursday?

(l) How much more money did the camp receive on Saturday than
on Wednesday?

(m) Some sites were rented for a single night and others for multiple nights.
If the camp received $905 for single night rentals on Friday, how many
sites were rented for multiple nights that night?

6 There are 540 angelfish in a pet store.
There are 5 times as many goldfish as angelfish.
How many fewer angelfish are there than goldfish?

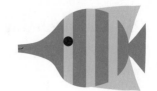

7 A florist had 285 roses.
19 of them wilted.
She wants to make bouquets of 3 roses each with the remaining roses.
How many bouquets can she make?

8 A teacher bought 32 notebooks at $3 each.
He paid for the notebooks and got $4 change.
How much money did he give the cashier?

NOTE
BOOK

198 pages / 3" × 5"
Recycled Materials

9 A roll of ribbon 315 ft long is cut into 3 pieces,
A, B, and C.
B is 25 ft longer than A.
C is twice as long as B.
How long is A?

10 8 lamp posts are an equal distance from each other along a street.
The distance from the 2nd to the 6th lamp post is 500 ft.
What is the distance from the 3rd to the 8th lamp post?

Exercise 4 • page 196

220   Review 2